T0251336

SWISS COMPETENCES IN RIVER ENGINEERING AND RESTORATION

SPECIAL SESSION ON SWISS COMPETENCES IN RIVER ENGINEERING AND RESTORATION OF THE SEVENTH INTERNATIONAL CONFERENCE ON FLUVIAL HYDRAULICS (RIVER FLOW 2014), EPFL LAUSANNE, SWITZERLAND, 5 SEPTEMBER 2014

Swiss Competences in River Engineering and Restoration

Editors

Anton J. Schleiss
École Polytechnique Fédérale de Lausanne (EPFL), Switzerland

Jürg Speerli
Hochschule für Technik Rapperswil (HSR), Switzerland

Roger Pfammatter
Schweizerischer Wasserwirtschaftsverband (SWV)/Association Suisse pour l'aménagement des eaux (ASAE), Switzerland

CRC Press
Taylor & Francis Group
Boca Raton London New York Leiden

CRC Press is an imprint of the
Taylor & Francis Group, an **informa** business

A BALKEMA BOOK

Cover photo: "Alter Rhein" River near Diepoldsau (SG), Switzerland. Internationale Rhein-regulierung (IRR)

CRC Press/Balkema is an imprint of the Taylor & Francis Group, an informa business

© 2014 Taylor & Francis Group, London, UK

Typeset by V Publishing Solutions Pvt Ltd., Chennai, India
Printed and bound in Great Britain by CPI Group (UK) Ltd, Croydon, CR0 4YY

All rights reserved. No part of this publication or the information contained herein may be reproduced, stored in a retrieval system, or transmitted in any form or by any means, electronic, mechanical, by photocopying, recording or otherwise, without written prior permission from the publisher.

Although all care is taken to ensure integrity and the quality of this publication and the information herein, no responsibility is assumed by the publishers nor the author for any damage to the property or persons as a result of operation or use of this publication and/or the information contained herein.

Published by: CRC Press/Balkema
P.O. Box 11320, 2301 EH Leiden, The Netherlands
e-mail: Pub.NL@taylorandfrancis.com
www.crcpress.com – www.taylorandfrancis.com

ISBN: 978-1-138-02676-6 (Hbk + CD-ROM)
ISBN: 978-1-4987-0443-4 (eBook PDF)

Swiss Competences in River Engineering and Restoration – Schleiss, Speerli & Pfammatter (Eds)
© 2014 Taylor & Francis Group, London, ISBN 978-1-138-02676-6

Table of contents

Conference papers

Preface

The world is like a river, running along in its bed, this way and that, forming sandbanks by chance and then being forced by these to take a different course. Whereas this all proceeds smoothly and easily and gradually, the river engineers have great difficulties when they seek to counteract this natural behaviour (Goethe).

Goethe recognized that the dynamics of a river can only be controlled to a limited extent by channel modifications and rigid river training works. The term "dynamics" refers to variations in hydromorphology over space and time due to flood discharges and sediment transport. These processes regularly lead to the destruction of habitats, especially in riparian areas, and the creation of space for new habitats. Dynamic watercourses require a lot of space. For example, naturally meandering rivers may migrate laterally within a belt of roughly 5–6 times the width of the channel bed. In the valleys of the Swiss Alps and Pre-Alps the rivers originally divagated over the entire valley floor.

To reclaim land for urban development and agriculture as well as to provide flooding, watercourse alterations were carried out over the last two centuries in Switzerland. Efforts were thus made to impede the dynamics; rivers and streams were channelized, and channel bed widths were optimized with regard to sediment transport. This resulted in monotonous watercourses, with almost no variation in hydraulic or morphological characteristics.

Recognizing the ecological deficit of the Swiss, a new approach in the strategic planning of flood protection projects was promoted by the Swiss Government which gave the basis for the first restoration programs more than 40 years ago. Since then much has been achieved. Nevertheless, today's challenge of river engineers, in collaboration with environmental scientists, is to restore the channelized rivers under the constraints of high urbanization and limited space. The behaviour of river systems is a result of the complex interaction between flow, sediments, morphology and habitats. Furthermore, rivers provide important sources of water supply and energy production in addition to a means of transportation.

Each year the Swiss Commission for Flood Protection (KOHS) of the Swiss Association for Water Management (SWV) organizes a symposium where professionals, officers of public administrations and researchers exchange their experiences on special topics and on-going projects. In 2014 this symposium was organized as a special session of the seventh International Conference on Fluvial Hydraulics "River Flow 2014" at École Polytechnique Fédérale de Lausanne (EPFL), Switzerland. Aside from the Swiss participants, scientists and professionals from all over the world were informed about the Swiss competences in river engineering and restoration. In the presented book, invited and selected contributions regarding the latest tendencies and key-projects in Switzerland are presented to an international community of river engineers and researchers, hoping that they can enrich flood protection and river restoration projects all over the world.

We acknowledge the support of the Swiss Federal Office for the Environment, BG Consulting Engineers, and Hydro Exploitation SA as main sponsors for the proceedings and the River Flow 2014 conference. Further support to the conference and the special Swiss session was given by the following sponsors: e-dric.ch, IM & IUB Engineering, Basler & Hofmann, Met-Flow SA, and AquaVision Engineering; as well as Stucky, groupe-e, Patscheider Partner, HydroCosmos SA, Kissling + Zbinden AG, Ribi SA, and Pöyry. Finally, we recognize the support of the Laboratory of Hydraulic Constructions (LCH) of EPFL and the Swiss Association for Water Management (SWV), which organized the special session «Swiss Competences in River Engineering and Restoration» in the framework of River Flow 2014.

Prof. Dr. Anton J. Schleiss, *Conference Chairman of River Flow 2014*
Prof. Dr. Jürg Speerli, *Chairman of the Swiss Commission on Flood Protection (KOHS)*
Roger Pfammatter, *Director of the Swiss Association for Water Management (SWV)*

Swiss Competences in River Engineering and Restoration – Schleiss, Speerli & Pfammatter (Eds)
© 2014 Taylor & Francis Group, London, ISBN 978-1-138-02676-6

Organization

MEMBERS OF THE INTERNATIONAL SCIENTIFIC COMMITTEE

Walter H. Graf, Switzerland *(Honorary Chair)*

Jorge D. Abad, *USA*
Jochen Aberle, *Norway*
Claudia Adduce, *Italy*
Mustafa Altınakar, *USA*
Christophe Ancey, *Switzerland*
William K. Annable, *Canada*
Aronne Armanini, *Italy*
Francesco Ballio, *Italy*
Roger Bettess, *UK*
Koen Blanckaert, *China*
Robert Michael Boes, *Switzerland*
Didier Bousmar, *Belgium*
Benoît Camenen, *France*
António Heleno Cardoso, *Portugal*
Hubert Chanson, *Australia*
Qiuwen Chen, *China*
Yee-Meng Chiew, *Singapore*
George S. Constantinescu, *USA*
Ana Maria da Silva, *Canada*
Andreas Dittrich, *Germany*
Rui M.L. Ferreira, *Portugal*
Massimo Greco, *Italy*
Willi H. Hager, *Switzerland*
Hendrik Havinga, *The Netherlands*
Hans-B. Horlacher, *Germany*
David Hurther, *France*
Martin Jäggi, *Switzerland*
Juha Järvelä, *Finland*
Sameh Kantoush, *Egypt*
Katinka Koll, *Germany*
Bommanna G. Krishnappan, *Canada*
Stuart Lane, *Switzerland*
João G.A.B. Leal, *Norway*
Angelo Leopardi, *Italy*

Danxun Li, *China*
Juan-Pedro Martín-Vide, *Spain*
Bijoy S. Mazumder, *India*
Bruce W. Melville, *New Zealand*
Emmanuel Mignot, *France*
Rafael Murillo, *Costa Rica*
Heidi Nepf, *USA*
A. Salehi Neyshabouri, *Iran*
Vladimir Nikora, *UK*
Helena I.S. Nogueira, *Portugal*
Nils Reidar Olsen, *Norway*
André Paquier, *France*
Piotr Parasiewicz, *USA*
Michel Pirotton, *Belgium*
Dubravka Pokrajac, *UK*
Sebastien Proust, *France*
Wolfgang Rodi, *Germany*
Jose Rodriguez, *Australia*
Pawel M. Rowinski, *Poland*
André Roy, *Canada*
Koji Shiono, *UK*
Graeme M. Smart, *New Zealand*
Sandra Soares-Frazão, *Belgium*
Thorsten Stoesser, *UK*
Mutlu Sumer, *Denmark*
Simon Tait, *UK*
Aldo Tamburrino, *Chile*
Wim S.J. Uijttewaal, *The Netherlands*
Zhaoyin Wang, *China*
Volker Weitbrecht, *Switzerland*
Silke Wieprecht, *Germany*
Farhad Yazdandoost, *Iran*
Yves Zech, *Belgium*

MEMBERS OF THE LOCAL ORGANIZING COMMITTEE

Anton J. Schleiss, *EPFL, Conference Chairman*
Giovanni De Cesare, *EPFL, Co-Chairman*
Mário J. Franca, *EPFL, Co-Chairman*
Michael Pfister, *EPFL, Co-Chairman*
Scarlett Monnin, *EPFL, Secretary*
Gesualdo Casciana, *EPFL, Secretary*

Swiss Competences in River Engineering and Restoration – Schleiss, Speerli & Pfammatter (Eds)
© 2014 Taylor & Francis Group, London, ISBN 978-1-138-02676-6

Sponsors

SUPPORTING INSTITUTIONS

 Schweizerischer Wasserwirtschaftsverband
Association suisse pour l'aménagement des eaux
Associazione svizzera di economia delle acque

SWISS NATIONAL SCIENCE FOUNDATION

ÉCOLE POLYTECHNIQUE
FÉDÉRALE DE LAUSANNE

GOLD SPONSORS

 Schweizerische Eidgenossenschaft
Confédération suisse
Confederazione Svizzera
Confederaziun svizra

Swiss Confederation

Federal Office for the Environment FOEN

SILVER SPONSORS

Basler & Hofmann

 MET-FLOW

IM Engineering | **IUB** Engineering

AquaVision Engineering

KOHS SPONSORS

patscheiderpartner
E N G I N E E R S

ribi
sa ingénieurs
hydrauliciens

 PÖYRY

 HydroCosmos

 groupe e

stucky >
a Gruner company

Invited papers

Restored Aire River near Geneva (GE), Switzerland. Photo: Bernard Lachat, Biotec.

Swiss strategy for integrated risk management: Approaches to flood protection and river restoration

J. Hess, O. Overney & E. Gertsch
Federal Office for the Environment FOEN, Switzerland

ABSTRACT: the Swiss strategy for integrated risk management is based on a holistic approach seeking the optimal combination of response, recovery and preparedness. This approach has been developed in a strategic report in 2011 into 6 priorities for action: comprehensive knowledge of hazards and risks (1); increased awareness of natural hazards (2); holistic planning of measures (3); protective structures designed to accommodate excess loads (4); emergency preparedness (5); timely identification of hazard events (6). The implementation of these priorities is illustrated with examples from different flood risk management projects in various basins in Switzerland.

1 INTRODUCTION

Switzerland has a long history and experience in dealing with natural hazards. However, only in 1987 in the aftermath of major floods, it became clear that structural measures alone are not sufficient to guarantee protection. Since then spatial planning (master planning and land-use planning) has obtained far greater priority in the context of sustainable and hazard-conscious land use. The idea that sufficient space must be given to watercourses also became accepted.

Recent events in the years 1993, 1999, 2000, 2005 and 2007 also showed that damage could be significantly reduced with the help of modern protection concepts: robustly designed protection structures that are conceived to cope with excess loads are the key factors for successful prevention (Bezzola & Hegg, 2007). Moreover, the damage caused by floods can be reduced by around one fifth if the authorities issue timely warnings and alerts and people takes suitable measures to protect their lives and property as part of their own individual responsibility.

In 1997, the Federal Council founded the National Platform for Natural Hazards, PLANAT, an extraparliamentary commission of the Federal Department of the Environment, Transport, Energy and Communications (DETEC), with the aim of preventing an increase in the damage caused by natural hazards, providing sustainable protection for living space and improving hazard prevention. PLANAT was then commissioned by the Federal Council to develop the strategy entitled Sicherheit vor Naturgefahren ("Safety against Natural Hazards") (PLANAT, 2004). The aim of this strategy is to provide a comparable level of safety throughout Switzerland for all natural hazards that would be ecologically justifiable, economically viable and socially responsible. The PLANAT strategy raises awareness of a risk-based philosophy and promotes integrative risk management in the area of natural hazards.

2 PRINCIPLES OF INTEGRATED RISK MANAGEMENT

2.1 Risk management approach

The tasks involved in risk management consist in the continuous monitoring of the relevant factors and periodic recording of the relevant risks (see Figure 1). Risks should be assessed in relation to their acceptability. The action required and priorities for the steering of future developments through the implementation of suitable measures are deduced from this information. Through suitable measures, new unacceptable risks are avoided, unacceptable risks alleviated and acceptable risks borne. Intensive risk dialogue between all actors is a precondition for effective risk management. (PLANAT, 2013).

Risk management provides answers to three key questions (see Table 1):

2.2   Integrated flood Risk Management (IRM)

An integrated and holistic risk management assumes that all types of measures for natural disaster reduction are considered. Generally, measures of preparedness, response and recovery (reconstruction) are equally applied. The measures for dealing with natural hazards cover the three phases of the risk cycle (see Figure 2).

Planning flood protection works need to integrate both ecological and security aspects. All measures must complain with sustainability and must provide a good cost-benefit relation.

Integrated flood risk management deals, on one side, with the natural hazard processes and, on the other side, with damages and risks. Sound scientific knowledge in hydrology

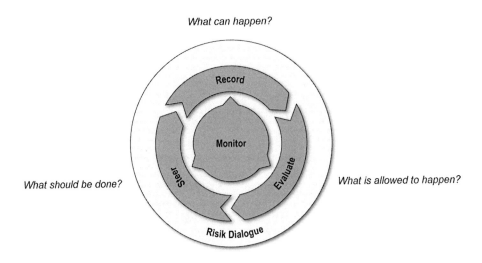

Figure 1.   Risk management approach (Source: PLANAT, 2013).

Table 1.   Three key questions of risk management (Source: PLANAT, 2013).

Question	Answer
What can happen?	The risk analysis is based on systematic and scientific processes. Both the intensity and frequency of natural hazards and the damage to be expected are recorded.
What is allowed to happen?	As part of the risk assessment it is decided which risks are considered acceptable and unacceptable. A risk that is assessed as bearable for good reasons is an acceptable risk.
What should be done?	Future risks are maintained at an acceptable level, existing risks are reduced to an acceptable level and the approach to residual risk is managed through the implementation of measures. Integrative action planning is an optimization process, in which risks and opportunities are weighed up, and proportionality must be maintained in relation to all dimensions of sustainability. The extent to which risks should be avoided, reduced and borne is also decided as part of this process.

and hydraulic are fundamental to evaluate correctly flood hazards. Access to information of land use planning and to insurance data is also necessary for the evaluation of vulnerability and resilience. Only with sufficient appropriate data, flood risk management will achieve an optimal use of all chances to influence hazards and risks.

2.3 Implementation in a federal state

As Switzerland is a federal state, the institutional implementation of integrated risk management is based on the delegation of competence at different levels (see Table 2). The responsible actors are involved due to a legal obligation or because they assume individual responsibility. Subsidiarity plays an important role as a principle of delegation.

The federal state defines the strategy and the legislative framework, the cantons and the municipalities implement the strategy through land use planning, as well as maintenance and

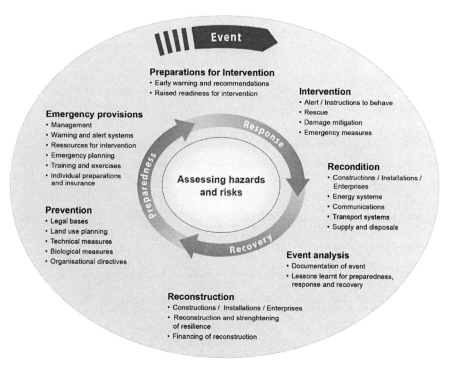

Figure 2.   Range of measures involved in integrative risk management and phases in which they are implemented (Source: BABS, 2012).

Table 2.   Task sharing in flood risk management.

Level	Tasks
Federal authorities	Legislation, policy, guidelines, financial support, support of research, education, warning and alerting
Cantons (26)	Enforcement of laws, cantonal structure planning, hazard mapping, planning of measures on regional scale, cantonal emergency management
Municipalities (2408)	Communal land use planning, building permissions, planning of measures on local scale, local emergency management
Property owner	Local protection, precautionary measures
Insurance	Mandatory insurance (all buildings), covering the remaining risk

construction of flood protection works. The federal state support hazard mapping and flood protection measures through financial subsidies. Property owners and insurances play an important role, as they have to bear residual risks through flood proofing or compensation.

3 INTEGRATED FLOOD RISK MANAGEMENT WITH EXAMPLES OF ACTUAL FLOOD PROTECTION PROJECTS

3.1 Actual flood protection projects

Many flood protection works in Switzerland were realised in the 19th century and need to be completely rehabilitated. Recent flood events in the last 30 years (1987, 1999, 2000, 2005, 2007) have shown that design values should be revised to take into account statistic uncertainties and climatic evolution. Hydraulic capacities that were designed at the beginning of the 20th century are not sufficient to ensure contemporary safety standards, moreover with the development of higher potential damages.

As a result, flood protection works along many rivers in Switzerland are being rehabilitated in order to match modern standards (see Figure 3). The most important actual flood protection project is the third correction of the Rhone River. Another important flood issue is the flooding risk due to the Sihl in the city of Zurich. Flood protection rehabilitation projects are also undergoing on all important alpine and pre-alpine rivers (Aare, Reuss, Linth). These projects can illustrate the way flood protection issues are tackled in Switerzland and how actions are undertaken to reduce flood risks.

3.2 Priorities of action

In 2011 the Federal Office for the Environment FOEN has defined 6 priorities for action in a strategy paper on "living with natural hazards" (FOEN, 2011):

1. Comprehensive knowledge of hazards and risks
2. Increased awareness of natural hazards

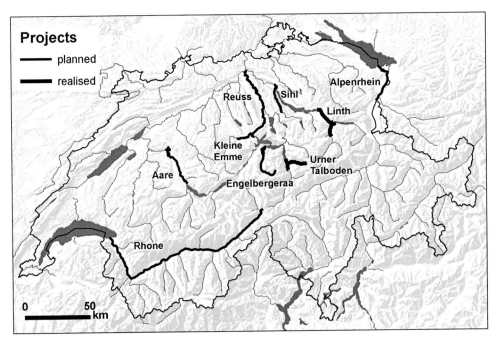

Figure 3. Major projects of flood protection in Switzerland.

6

3. Holistic planning of measures
4. Protective structures designed to accommodate excess loads
5. Emergency preparedness
6. Timely identification of hazard events

In the following chapters, the 6 priorities will be explained theoretically. Four of these priorities for action will be illustrated practically with the example of the third correction of the Rhone River.

3.2.1 *Comprehensive knowledge of hazards and risks*

Central to the integrated flood risk management cycle are hazard and risk assessments. A society can only deal sensibly with natural hazards if it has an in-depth knowledge of the hazards, assesses them objectively, takes preventive measures and reacts quickly and correctly in the case of an emergency. Therefore, hazard fundamentals (incl. event analysis to support economic viability for resilience building) are of primary importance for effective and efficient flood risk management. Hazard assessment is relevant to determine the magnitude and frequency of environmental processes in affected areas, taking into account already existing protective structures. The result of the hazard assessment is represented in a hazard map. The results of assessments and simulations are compared with the records of previous natural hazard triggered disasters.

Hazard maps are the usual tool for implementation of flood risk management in land use planning in order to avoid new zoning in endangered areas. Hazard maps are not always adequate for prioritisation of flood management measures. For example in the case of Zurich city, the hazard level due to flooding from the Sihl and the Limmat rivers is relatively low (<50 cm) but this does not reflect correctly the risk level. In fact the vulnerability shown by the density of employment gives a much more precise picture of the risk (see Figure 4) and their economic significance. The urgent need to address this flooding risk allows to prioritize the measures and to define an order for their implementation (see 3.2.6).

3.2.2 *Increased awareness of natural hazards*

The events of recent years have shown that the population is not very familiar with natural hazards. However, it is important to conserve and promote the knowledge that already exists

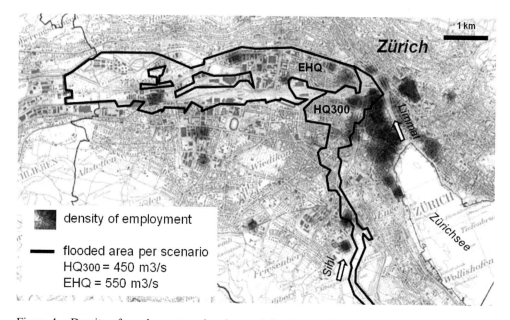

Figure 4. Density of employment and endangered flood zones in the city of Zürich (dark grey = 250 jobs/ha), (Source: Baudirektion Kanton Zürich, 2013, modified).

on dealing with natural hazards and to document and analyse new hazard events so that lessons can be learned from them. The population should be informed in a targeted way and on all levels about the relevant natural hazards. This process starts at primary school level through the setting of corresponding learning goals for geography classes. These conditions are essential to enable the establishment of a risk dialogue. Good hazard event documentation and other easily accessible information should ensure that the population does not forget existing hazards and assumes greater personal responsibility for hazard mitigation. This requires the provision of solid basic training in natural hazards for all those involved in the planning and construction of buildings, facilities and infrastructure as knowledge about the vulnerability of buildings is crucial to the minimisation of damage. International cooperation, as well as research and development in the field of natural hazards are also important components of natural hazard knowledge management. (FOEN, 2011).

3.2.3 *Holistic planning of measures*

The principle of flood risk management is the optimal combination of structural, biological, land-use planning and preparedness measures along with insurance protection. Whereas comprehensive hazard fundamentals are central to the approach and preparedness, response and recovery are the main complementary parts.

In the phase before an incident, measures of prevention and mitigation and measures to cope with an incident (preparedness) are taken. Prevention pays out. Investment in flood risk reduction protects lives and livelihoods, public assets and private property. It pays off on a major scale through minimizing the vulnerability of people and material assets to natural hazards. On the one hand damage is primarily avoided by an appropriate land-use planning based on hazard and risk mapping. Where it is not possible to avoid hazards structurally, technical measures (dikes, dams, etc.) or biological measures (silvicultural and eco-engineering measures) have to be taken, which are supposed to minimize the intensity of the hazard. On the other hand damage is avoided by managing and coping with the disaster. Preparedness measures are provisions for emergency situations that can occur and must be managed. Examples of such organizational measures are the implementation of warning systems, emergency intervention and rescue planning, training and public simulation exercises or insurance purchasing for house owners etc.

Because of insufficient hydraulic capacity and high risk of dyke failure, the profile of the Rhone River must be entirely new designed. The main constraint is not to enhance the dyke height and therefore the water level during flood event. The riverbanks should be large, not steep, so that protective works against side erosion are simple, robust and adaptive. A riparian vegetation can grow on these banks and contribute to the bank stability and to the biodiversity. All in all the Rhone River bed should be widened for 60%, which implies an augmentation of the river corridor of 870 hectares for an actual surface of 1380 ha (see Figure 5). The redesign of the river through systematic dam elevation was dismissed. A hydraulic analysis has shown that dam elevation was not a robust solution because the water level would rise higher during extreme event and so increase residual risk due to dam break. The flood plain would face the same hazard than today but on a much higher level and lower probability. Moreover raising the dams has negative consequences for the groundwater level and makes the drainage of the floodplain almost impossible. Finally this solution is not sustainable as it offers very limited possibilities for later adaptations.

3.2.4 *Protective structures designed to accommodate excess loads*

A lesson learnt from previous flood events in the Alps is the possibility of events of much higher magnitude than the design value used for protection work. As we cannot afford to design our works for all possible magnitude or process, we try to take into account an overload case in the design of our protection systems. The first goal is to avoid uncontrolled collapse of the protective works and the second is to handle the overload with non-structural measure.

The Engelberger Aa River project is designed to deal with extreme events well above the design value of the dyke. Through a combination of flood routing and flood diversion

Figure 5. Comparison of the old (top) and new (bottom) planned profile for the Rhone River (Source: Etat du Valais, 2008, modified).

measures extreme floods are conducted in flood corridors. Although damages are expected to occur during extreme floods, they can be reduced if only one side of the valley is flooded. This principle has been implemented on the section of the Engelberger Aa in Nidwald, where potential damages of the last section of valley are highest but the hydraulic capacity is limited and can not be enhanced due to the urbanization.

3.2.5 Emergency preparedness

Careful emergency planning helps to reduce the damage caused by extreme natural hazard events. Communes must have an emergency concept, based on natural hazard information sources, and regularly rehearse the measures necessary for the successful management of such events. Expert support is provided by national and cantonal agencies (documentation and training). Since its reform in 2004, civil protection in Switzerland has been organised as a civil network involving the cooperation of five partner organisations: i.e. police, fire brigades, health system, technical operations and civil protection. They ensure that management, intervention, protection, rescue and assistance services are provided in the course of extraordinary situations. Good cooperation between experts, management and emergency personnel is crucial here.

Overall responsibility lies with the cantons, however the main responsibility for emergency planning and organisation lies with the communes. In addition, the federal authorities may fulfil coordination or management tasks in the case of major hazard events in agreement with the cantons (Federal Act on Civil Protection and Civil Defence). The Federal Office for Civil Protection (FOCP) supports the cantons and partner organisations through its agencies (e.g. National Emergency Operations Centre) in the areas of planning, coordinating and implementing of civil protection intervention and emergency measures. If the civil protection and defense resources prove insufficient, military resource may also be made available to support the civil management bodies (subsidiary deployment of the army). (FOEN, 2011).

3.2.6 Timely identification of hazard events

Damages can only be limited if timely action can be taken at local level. This necessitates the perfect functioning of forecasting and warning chains and the interpretation of the available information at the end of this chain through on-site observations in the local context.

Planning of flood protection projects can take several years before protection works are built and offer an effective protection. Participatory planning processes and construction

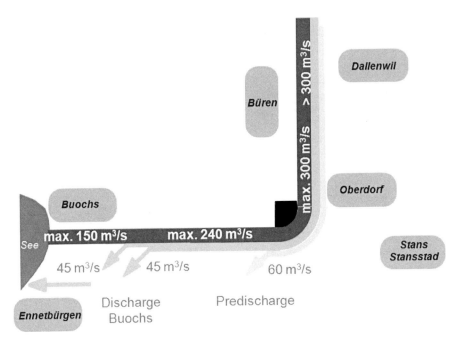

Figure 6.   Overload case management Engelberger Aa (Source: Kanton NW, 2009, modified).

permit proceedings are quite complex to manage and need time and energy from the planning team. It is however expected from the authority to take actions as soon as the hazard situation is identified. Because forecasting and warning systems can be implemented more quickly than protective works, they are very often used as anticipated measure to reduce risk in a transition phase. For example the protection measures in the city of Zurich will need an exhaustive planning process before the optimal alternative for flood protection is realised and effective. Therefore the canton has established a forecasting and warning system combined with a detailed emergency planning including all partners (railway, fire brigade, etc.).

A detailed hydrologic study has demonstrated that the artificial lake in the upper part of the Sihl basin could help reduce the peak flow during an extreme event and so contribute to reduce risks. In collaboration with the hydroelectric company that manages the lake and which is a subsidiary corporation of the railway company, the canton has set a flow and level regulation plan for emergency situation based on a hydro-meteorological forecast model.

4   CONCLUSION

An efficient flood risk management can only be achieved if all possible measures are effectively taken by a clear division of tasks between public authorities. Responsibilities must be clarified between the different state levels and the private sector (insurance companies and property owner). In addition good cooperation is crucial to the fulfilment of the stated objectives.

The successful implementation of integrative risk management coordinates the action priorities: protective structures alone cannot guarantee safety. An optimal combination of response, recovery and preparedness measures must be sought under financial, social and ecological constraints.

The tasks to be carried out to attain the defined objectives are known. The most recent hazard events confirm the appropriateness and efficiency of the hazard protection strategy and show that the approach that has been adopted is the correct one. The task now is to ensure the consistent implementation of integrative risk management. (FOEN, 2011).

REFERENCES

BABS (Swiss Federal Office for Civil Protection). 2012. Kreislauf des integralen Risikomanagements. http://www.bevoelkerungsschutz.admin.ch/internet/bs/de/home/themen/gefaehrdungen-risiken. html, 8.5.2014.

Baudirektion Kanton Zürich, 2013: Hochwasserschutz an Sihl, Zürichsee und Limmat: Integrales Risikomanagement und Massnahmenziel – Konzept. Bericht, Dezember 2013.

Bezzola, G.R. & Hegg, C. (Ed.). 2007. Ereignisanalyse Hochwasser 2005. Bundesamt für Umwelt BAFU, Eidgenösische Forschungsanstalt WSL.

Etat de Valais. 2008. Rapport de synthèse du plan d'aménagement de la 3e correction du Rhône. Mai 2008.

FOEN (Federal Office fort he Environment. 2011. Living with Natural Hazards.

Kanton NW, Tiefbauamt, 2009: Integrales Risikomanagement am Beispiel Engelberger Aa. März 2009

PLANAT. 2004. Sicherheit vor Naturgefahren—Vision und Strategie. Nationale Plattform Naturgefahren PLANAT.

PLANAT. 2013. Sicherheitsniveau für Naturgefahren. Nationale Plattform für Naturgefahren PLANAT.

Swiss Competences in River Engineering and Restoration – Schleiss, Speerli & Pfammatter (Eds)
© 2014 Taylor & Francis Group, London, ISBN 978-1-138-02676-6

Revitalization of rivers in Switzerland—a historical review

Christian Göldi
Schaffhausen, Switzerland

ABSTRACT: Most rivers in Switzerland were canalised and many small creeks have been put in culverts during the last 150 years to protect villages and infrastructure against floods and to enlarge agricultural land. The natural character of these watercourses changed therefore to a monotonous and boring shape. The loss of the habitats with their great variety of animals and plants aroused a fundamental change in the way how to implement river construction works. Nature friendly techniques like construction methods with living plants (willows and others) were developed and tested. The new approach leads to an astonishing increase of ecological values. Nature friendly river works and revitalization of canalised rivers is now part of the federal law in Switzerland.

1 WATER CONSERVATION AT THE BEGINNING

In the late 60ies the main issue was the enormous water pollution in the lakes and rivers of Switzerland. It was first an immense technical and political achievement of the country, to solve that environmental problem. The water quality in our rivers and lakes is now generally good. There are still some problems to solve. The contamination with active hormone parts and the infiltration parts of agricultural fertilizer and herbicides was not reduced well enough since. To save the groundwater in the future, it will be necessary to state more protected areas. The concepts of water conservation was the collection of sewage water in pipes connected to sewage treatment plants. The so cleaned water was discharged into the "Vorfluter" (The German word Vorfluter is a simplifying and disrespectful expression for the receiving river!).

At that time, the "Vorfluter" was the end of the Process. The planners did not go further. They did not care what comes after. The head of the department for water conservation of the Canton Zürich stated in 1980: "It is a frustrating mind to see the product of excellent purified sewage water disappear in a canalized or covered water course". It is no surprise, that he strongly supported new natural construction measures for river works and later initialized the Revitalization Program for Rivers in his canton.

2 RIVERS IN THE LANDSCAPE

Rivers are dangerous, rivers cause a lot of damages with their floods, rivers must be tamed. In the 19th century, population of Switzerland has grown from 1.7 Million to 3.5 Million. The demand of more land for agriculture, for houses and infrastructure was immense.

In the last 200 years most large rivers were straightend and narrowed between embankments. The danger of floods was significantly reduced. Protected against floods, infrastructure facilities like railway lines and roads could be build. Towns and cities could expand their living, comercial and industrial areas. The plains were transformed into fertile farm land. To gain additional agricultural areas, wetlands have been drained and a great number of small waterways have been put into culverts. The country was proud of its deeds like the correction projects Kanderdurchstich, Linth Project, the Limmat, the Rhine, the Juragewässerkorrektion etc. Water engineers of today think with great respect of the big schemes of river works in the past. Nevertheless today we realise, that with these works the landscape has changed her original situation in a more structured geometrical shape. Lots of natural spots, dynamic rivers and easy meandering creeks got lost. Living space for a great variety of plants and animals disapeared dramatically.

2.1   *Rhäzünser Auen*

The parliament of Switzerland (Bundesversammlung der Schweizerischen Eidgenossenschaft) decided (act of Parliament) in 1960 the program for the national highway system. The route between Germany and Italy over the San Bernardino Pass in the Alps is part of the system. To find an easy layout of the road, at that time it was obvious that river areas can be used. The planners drew a project straight through the meandering river Hinterrhein and its alluvial forest between Rhäzuns and Rotenbrunnen. 1968 the project was approved by the authorities of Canton Graubünden. In this sector the river was very natural and untouched by technical measures, like dams or cascades. It was the last remaining natural stretch with a length of about 3 kilometers. On the overall length from the gorge of Viamala, where the River Rhine leaves the narrow Alps and enters into the plains right down to the Lake of Constance, the river has been regulated and straightened more than 100 years before. National organisations, engaged for nature protection, then have not been aware what ruinous impact this project would cause to the beautiful fluvial landscape. Only a few persons realised the dramatic situation. The cantonal officer for landscape preservation alarmed the federal engineer in charge for road construction. Both tried to convince their superiors to modify the project into an environmentally acceptable solution with tunnels. Surprisingly in 1975 the federal counsel decided against the government of Canton Graubünden. To the astonishment of the people, the highway has be placed underground into a tunnel. They concluded that this beautiful wild landscape with relevance to the whole of Switzerland must be protected and preserved in its natural state. The alternative project has been built in the period 1981 to1983.

The story of Rhäzünser Auen is one early case where nonmonetary values outreached technocratic and financial arguments. It was a signal for the protection of natural rivers and

Figure 1.   Project 1968 (montage).

Figure 2.   Rhäzünser Auen today (Ch. Göldi).

Figure 3.   Reussdelta (O. Lang).

14

as we see it today as a strong support for river works close to nature and the revitalization of rivers.

2.2 Reussdelta project Vierwaldstättersee (lake of Lucerne) Canton Uri

The Reuss River in the center of Switzerland has been canalised within the plains of the Canton Uri in the middle of the 19th century. The estuary into the Vierwaldstättersee has been prolonged to allow a better transport of sediment coming from the mountains. Since 1905 gravel was exploited in the delta. Problems with the stability of the borders of the lake and the dissatisfying ecological situation induced the authorities of the Canton Uri 1970 to establish a commission of experts to develop a delta project, which considers all interests of ecology, fisheries, landscape, recreation and exploitation of gravel. The studies and modelling in the Laboratory of Hydraulics, Hydrology and Glaciology (VAW) at the Swiss Federal Institute of Technology gave a strong support for the development of the project.

The project was a new approach in river works. 1988 machines opened the banks on both sides of the canal to allow the Reuss River to develop a new natural River. It is the first revitalization of a canalised river mouth and delta in Switzerland.

3 THE ORIGIN OF RIVER REVITALIZATION BASES ON THE IDEAS TO IMPLEMENT CLOSE TO NATURE TECHNIQUES FOR RIVER WORKS

3.1 Bioengineering instruction course at Innsbruck with Prof. Schiechtl

At the Swiss Federal Office for Environment it was known that in Austria Prof. Schiechtl in Innsbruck has developed an alternative construction technique to consolidate hang slides, road dams and river embankments without concrete or steel. Educated in natural science and expert in botany he knew the ability of living plants to stabilise loose earth material with their roots and to protect the surface with their branches. He successfully realised several bioengineering projects to fortify instable landslides and to create very natural river boundaries. The Swiss Federal Office for Environment organised in 1979 an instruction course with Prof. Schiechtl for federal and cantonal engineers in charge of flood control and environment project. The course encouraged many of them to use this close to nature method at their own projects. They got strong support by Helgard Zeh, the soil bioengineering pioneer in Switzerland. The method got a revival mainly for River works. It has to be mentioned, that in the Canton Solothurn, a river inspector made his first experiments with bioengineering methods at the Dünneren creek already in 1976. The little river presents itself today like a natural water course.

3.2 Association for soil bioengineering

Many professionals took interest in these new/old methods. They established contact to the German Society for Bioengineering (Gesellschaft für Ingenieurbiologie e.V.) and exchanged ideas and experiences. 1987 the annual assembly of the society took place in Zürich, where the leading international experts presented their themes. The group of friends for Bioengineering grew constantly, so that the Swiss specialists founded in 1989 their own association for bioengineering (Verein für Ingenieurbiologie).

3.3 Mülibach Saland

Close to nature construction methods for River projects are based on bioengineering methods. An early example for projects at the beginning of a new alternative and natural way to do in river works is the project of Mülibach Saland (a small creek) in the valley of Töss. The project has been realised in 1980. Willows and alders were used instead of heavy blocks along the river bed to protect the embankment against erosion.

Figure 4.   Emme (river pear) BE/SO (river management).

Figure 5.   Mülibach Saland ZH (Ch. Göldi).

3.4   *Emmebirne BE und SO (River pear)*

The state of the art to stabilise river beds against vertical erosion were usually massive traverse structures like sills built of concrete and heavy stones. The River Emme in the center of Switzerland every year lowered its bed level by vertical erosion. The authorities of the Cantons Bern and Solothurn were aware of problems of the sinking groundwater and the instability of the embankments. They asked the Laboratory of Hydraulics, Hydrology and Glaciology (VAW) at the Swiss Federal Institute of Technology and the Institute of Geography of the University of Bern to study the problems of the Emme River and to provide proposals for a good solution. The two institutes presented their study "Emme 2050" in 1989. They proposed a new idea for the stabilisation of the river course: the Emmenbirne (pear shape widening). Instead of keeping the river bed in its narrowed character, the bed had to be widened within the outside lying levees as far as possible. The hydraulic effect is stupendous. The widening reduces the tractive force and therefore the sediment deposits can stabilize the bed. Without questioning the flood control, a new dynamic natural river section can develop. The first sections were realised in 1991/92 and in 1995.

4   LAWS FOR RIVER PROTECTION IN SWITZERLAND

Switzerland has 26 Cantons (provinces) with their own constitution, parliament and government. The Cantons are responsible for water affairs. The Federal Government provides the legal framework and supports the effort of the Cantons for flood control and water protection financially with subsidies.

The way of the law of water protection to modern times in Switzerland is a long way. 1957 the first law for water protection was determined. 1975 the people of Switzerland voted with 77,5% yes for the citizens' initiative for adequate minimum water flow in the rivers (Wasserrechtsinitiative über die Sicherung angemessener Restwassermengen). The parliament had difficulties to find a solution for the law in time. A new citizen's initiative for the saving of rivers and lakes (Zur Rettung unserer Gewässer) was launched in 1984. It took another 8 years with controversial political discussions inside and outside the parliament until the positive vote in 1992 laid the basis for a nature- and landscape friendly law:

The main message is postulated in the *Swiss Federal Act on Water Engineering*, Article 4:

1. Waters, banks and flood protection structures shall be maintained in such a way that existing flood protection, and particularly the run-off capacity, is maintained.
2. In the event of interventions in bodies of water, the natural course of the water shall be retained or restored as far as possible. The water course and its banks shall be configured in such a way that:
   a) they are able to provide a habitat for a versatile world of flora and fauna
   b) the interplay between water above ground and water below ground is largely retained
   c) the appropriate bank vegetation for the location is able to flourish
3. In built-up areas, the authorities shall be empowered to approve exceptions to Paragraph 2.

It is remarkable that the same rule is postulated in the *Federal Act on the Protection of Waters Art. 37* and additional in Article 38 *1. Watercourses may not be covered or put in culverts.*

The advancement in laws concerning the water courses was not adequate for the Swiss Association of the Fishermen. They started 2005 a new citizen's initiative "living water" (Lebendiges Wasser). The initiative was well supported by the people. The federal parliament offered a counter proposal, so the initiative was withdrawn. 2011 the federal counsel could therefore introduced a new article in the act on the protection of Waters

Article 36a: contains the rule to *provide required space for rivers and lakes.*
Article 38a: explains how the cantons have to *rehabilitate their rivers and lakes.*

With all these legal regulations the way to improve and preserve the natural character of our rivers and lakes with its habitats of is well paved. First steps have been realized, as will be illustrated by the following examples.

5 CITY AND CANTON ZURICH: REVITALIZATION PROGRAMS FOR RIVERS

The Office of Water Protection and River Works Canton Zurich started 1983 first experiments for revitalization of canalised creeks.

5.1 *Nefbach at Neftenbach*

According to the project design of 1966, the Nefbach was built around 1978 in a hard technical way, a geometrical cross section, heavy rocks along the low water course, even the riverbed was fixed with a layer of big stones nicely placed without any structure.

The question was, if this canal, what was a lovely trout river in the past, has to stay in that sad form far into the future? One day in spring of 1983, a heavy machine was put into action. Without plan and mandate, the hard construction of the canal was broken on length of about 100 meter, the bed layer of fitted stones disturbed and the big stones of the bank protection newly placed in a rather wild arrangement. The result was astonishing. The canal has changed his character into a quite natural, diversified status. Fast flowing water zones were followed by slow current pools; deep sections and swallow areas allowed the sedimentation of variable sizes of material. The few trouts living in the canal immediately took over the more suitable biotope. And a few years later, when more sections were reshaped, the population of trouts grew very positive.

Figure 6.   Nefbach 1980 (agw).   Figure 7.   Machine on work 1986.

Figure 8.   Nefbach 1995 (Ch. Göldi, 7/8).

The experiment at the Nefbach convinced the chef of the department for water protection and his political superior. The idea for an all over revitalization program for canalised rivers was born.

5.2   *The parliament of Canton Zurich supports the program for revitalization of canalised rivers 1989*

The first successful experiments for revitalization at the Nefbach (1983/1986/1987) and at the Reppisch River (1985/1986) encouraged the staff of the section river maintenance, supported by their chef, to start a program for the revitalization of canalised rivers. They prepared a comprehensive documentation of the rivers and streams suitable for revitalization. When the minister for construction presented the program at the cantonal parliament, he explained that there not yet any federal law for river revitalization. He concluded that the implementation of measures for the revitalization of canalised and in culverts buried rivers is a task of high interest, which must be performed even without waiting for federal orders and laws.

The Kantonsrat of Zürich (parliament) voted without opposition for the approval of the program for the revitalization of rivers at his assembly at the 23rd of October 1989.

Since then until now, in the Canton of Zürich more than 90 km of earlier canalized rivers and creeks have been restored. Over 50 projects have been realized by the Canton and more than 300 by the communities. People enjoy the recreational impact of these open running rivers in their daily life, especially in urban areas and the improvement of natural habitats in the city and countryside. The population widely and strongly supports the principles of open and natural rivers. This was proofed at the plebiscite in 2005. The overwhelming positive vote for a new constitution of the Canton Zurich included a law to promote and support river revitalizations.

5.3   *Small rivers program in the City of Zürich*

The City of Zürich started his "Small Rivers Program" in 1988 before the canton was ready with his revitalization program for rivers and streams. Most small rivers in the city were put into culverts beginning more than a hundred years ago because of polluted water. These culverts were connected later to sewage treatment plants. The idea of the "Small Rivers

Figure 9. ZH Parliament 23rd October 1989 (Ch. Göldi).

Figure 10. Albisrieder Dorfbach Zürich.

Program" is, to separate fresh water running from the woody hills around Zurich and clean drain water of the urban area from sewage water. For the fresh water new rivers were built instead of conducting the water in culverts again. Since the start of the Small Rivers Program of the City more than 65 renewed small rivers are running open with a total length of over 25 kilometer.

6 SELECTED ACTIVITIES OF CANTONS

6.1 *Canton Basel-Landschaft und Canton Basel-Stadt—revitalization Birs River*

For more than 70 years, the Birs River flowed narrowed way between hard concrete embankments. There was little space for natural dynamic and wildlife habitat. In 2002 the two Cantons Basel-Landschaft and Basel-Stadt founded a project to revitalize the canalised River Birs. Today the Birs River is again a very natural flowing water course with gravel islands and varied structured river bounds. It has become again a habitat again for typical animals and plants. People enjoy the space for recreation. The two cantons were honoured for this initiative with the Swiss Water Award in 2007.

6.2 *Canton Aargau—natural park for floodplains*

1993 the people of the Canton Aargau voted with great approval for the citizen's initiative "Natural park for floodplains—support for endangered habitat". That voting expressed the awareness for the need of protection of the few left floodplains in the heart of Switzerland.

1989 big machines started to remove the hard bank protection measures and with the natural river dynamic more space was conquered. The river surroundings changed into attractive nature resorts and beautiful parts of the landscape.

6.3 Canton Bern

6.3.1 Pool for revitalization of rivers (renaturierungsfonds)

Following also citizens' initiative, in 1997 the people of Canton Bern clearly approved a cantonal article, within the law for water use and management, to establish a pool for financing ecological improvements of projects for large rivers and support revitalization projects for smaller rivers communities and private groups. The pool is supplied by 10 percent of the money the canton gets from the concessions of the hydro power producer companies.

Since the start more than 450 river projects have been realised and could be supported with over 28 Million Swiss Francs. Thanks to the pool of Canton Bern, the idea of river revitalization, got a real push.

In 2009 Canton Bern got the Swiss Water Award for the exemplary cooperation of his technical water administration and the section for nature rehabilitation for men and nature.

6.4 Canton Solothurn

6.4.1 Bioengineering at the Dünneren creek

As mentioned above, in Canton Solothurn, a river inspector made his first experiments with bioengineering methods at the Dünneren creek already in 1976.

Figure 11.   New River Aire 2002 GE (Ch. Göldi).

Figure 12.   River En near Samedan 2009 GR (Ch. Göldi).

6.4.2 Revitalization of in culverts buried and canalised rivers

The successful experiments with close to nature methods in the Canton Solothurn encouraged the water engineers to start revitalization projects. The first project was the opening of the Räberbächli in Matzendorf in 1985. More than 30 projects followed.

6.5 Canton Geneva

6.5.1 River revitalization as part of the law

The parliament of Canton Geneva changed the law for water policy in 1997 with the introduction of the obligation to revitalize the Rivers. The political preparedness to change unnatural technical parts of its rivers wherever it is possible with revitalization measures is very strong in Canton Geneva. The program for the revitalization of rivers launched in 1998 is an integrated part of the politics for construction activities of the canton. The Canton Geneva was the first winner of the Swiss Water Award in 2001.

6.5.2 Transboundary agreement with France (Le Contrat de Rivières transfrontalier entre Arve et Rhône)

Geneva borders mainly on France territory. The catchment areas most of its rivers (Rhone River excluded) is located partially in France. Quite a few rivers like the Aire are fed with water from France. To arrive a good partnership with the French communities near the border, the canton has developed a treaty which is called "Le Contrat de Rivières transfrontalier entre Arve et Rhône" to ensure solutions about water quality and flood control. The contract has been signed in October 2003.

Figure 13.   Thur River 1974 (agw).

Figure 14.   Thur River 2003 (Hj. Egger).

21

6.5.3 *The revitalization project of the river Aire*

The outstanding project of the revitalization program of Canton Geneva is the revitalization of the river Aire. The river has been canalised long ago. The first construction works on the site started in 2002. The works will continue until 2015. The planning team was honoured for this project by the Prix Schulthess in 2012.

6.6 *Canton Graubünden*

The Canton Graubünden and the city of Samedan was honoured for the outstanding River project En/Flaz by the Swiss Water Award 2005. Warned on earlier floods and a study of the flood risk, the authorities of the canton together with the counsel of the community of the city of Samedan, supported by the local population, developed a project which did not only reduce the danger of floods but also re-established the beauty of the landscape. The project En/Flaz gave the input to continue the revitalization works on the Inn River downstream. In a few years the valley of the Engadin will be enriched with a beautiful natural course of a renewed river.

6.7 *Canton Thurgau/Canton Zürich*

In 1978 a severe flood caused the overtopping and failure of the dams along the Thur River. Mostly agricultural land was flooded. Many ideas to protect the plains against the floods in the future aroused. It was a process to find a solution, which provides security and considers the demand of natural habitats and landscape. The river section in Altikon ZH/Niederneunforn TG is a successful example for revitalization and flood control over the border of two cantons.

7 CONCLUSION

The revitalization programs for rivers in Switzerland were and still are successful. The early experiments were very helpful to convince people and politicians in a vivid way directly on the site for the new approach how to work on rivers in a nature friendly way. Nature friendly river engineering is now widely accepted. One important reason for the successful realization of nature friendly river projects is the interdisciplinary team work between engineers, natural scientists, landscape planners and more specialists like agricultural people together with government officials and politicians. The early inclusion of NGO's and landowners into the project often helps to find good and widely accepted solutions. Nevertheless, the active support of engaged politicians is essential. Besides the revitalization of the rivers, protection against floods remains the main issue in river construction works. Such flood protection projects have to be implemented with high regard to landscape, nature, city planning and the people.

REFERENCES

Conradin F. & Räbsamen U. & Villiger J. 1988. Das Bachkonzept: Anlass, Ziel, Umfang. Sonderdruck: Das Bachkonzept der Stadt Zürich. Nr. 1164 aus *Gas-Wasser-Abwasser 1988/8* des Schweizerischen Vereins des Gas—und Wasserfaches, Zürich.

Göldi Ch. 1984 Naturnaher Wasserbau an Fliessgewässern—Ideen und Beispiele. *Gas Wasser Abwasser gwa 3, 113–121.* Göldi Ch. 2005. Der Wasserbau: Seit 1975 Teil des Umweltschutzes. *Zürcher Umweltpraxis Nr. 40, 47–50.* Baudirektion Kanton Zürich.

Gsell H.G. 1999. 10 Jahre Wiederbelebungsprogramm. *Gas Wasser Abwasser 11. 931–935.*

Renaturation du cours d'eau de l'Aire à Genève. 2012. Service de renaturation des cours d'eau, Genève.

Rentsch P. 1984. Ingenieurbiologie und ihre Anwendung im Gewässerunterhalt. *Schweizer Baublatt Nr. 68 24. August 1984.*

Willi H.P. & Göldi Ch. & Keller G. 1988. Kanton Zürich—Wiederbelebungsprogramm für die Fliessgewässer. Bericht, Direktion der öffentlichen Bauten des Kantons Zürich, Amt für Gewässerschutz und Wasserbau.

Swiss Competences in River Engineering and Restoration – Schleiss, Speerli & Pfammatter (Eds)
© 2014 Taylor & Francis Group, London, ISBN 978-1-138-02676-6

Swiss contribution to bed load transport theories

Martin N.R. Jaeggi
Jaeggi River Engineering and Morphology, Ebmatingen, Switzerland

Daniel L. Vischer
Laboratory of Hydraulics, Hydrology and Glaciology, Swiss Federal Institute of Technology, Zurich, Switzerland

ABSTRACT: Swiss rivers have been heavily trained during 19th century. In particular, the Alpine Rhine was narrowed and straightened by two cutoffs. The result was only partially satisfactory. This was the challenge for research on bed load transport for Eugen Meyer-Peter and his research team at the Laboratory of Hydraulics of the Swiss Federal Institute of Technology, opened in 1930, amongst them Hans Albert Einstein, Henry Favre, Charles Jaeger and Robert Mueller. A number of flume tests resulted in the so-called first Meyer-Peter formula (1935). More tests, including those performed in the so-called Meyer-Peter flume (2 m wide, 50 m long) as well as a thorough dimensional analysis led to the well-known Meyer-Peter/Mueller formula of 1948. Later, self-armouring process was in the focus of research led by Gessler and Guenter. Steep flume bed load transport tests allowed developing relations for bed load transport in mountain streams. Development of computer modelling allowed to assess the sediment regime of rivers; in particular that of the Alpine Rhine. Hunziker developed a procedure for fractionwise calculation of bed load transport. River restoration interests raised the question of sediment transport capacity in wide river reaches.

1 RIVER REGULATION IN SWITZERLAND

1.1 Early works

People for a long time avoided to settle in the river floodplains where they were under constant threat of floods. Increasing population and the need for more agricultural land built up the pressure on the alluvial plains. Some training works were undertaken in the middle age already. River regulations of the 18th and 19th century are described by Vischer (1986).

The first important project undertaken was the diversion of the Kander River to the Lake of Thun, realised between 1711 and 1714. Heavy sediment depositions of the Kander near Thun had gradually increased inundation risk. The main purpose of the project was to deposit the sediments in the lake. Works have been quite adventurous. A mountain ridge was crossed by gallery. The diversion induced heavy bed erosion, so that the gallery soon collapsed and the Kander now reaches the lake through a gorge.

1.2 The major regulations of the 19th century

Early 19th century, the Linth river was regulated (1807–1816). Increasing sediment deposits had turned the area between the lakes of Zurich and Walenstadt into a swampy zone, where malaria became endemic. The project led by Hans Conrad Escher consisted in a diversion of the Linth river into the Lake of Walenstadt and the building of a straight narrow incised channel between the two lakes, as well as a systematic drainage of the alluvial plain. The German engineer Johann Gottfried Tulla had a major impact on the project. Tulla (Mosonyi, 1970) later triggered the regulation of the Upper Rhine between Basle and Strassbourg, trained according to his principle, that 'no river needs more than one single channel!'.

On the model of these earlier works, most of the Swiss alluvial rivers have been regulated. The originally braided or meandering channels were forced into a narrow single thread channel. Fluvial hydraulics formulas like Chézy's were known and applied. Straightening and narrowing of channels increased bed load transport capacity. It might however seem surprising that all these heavy works have been realised before development of bed load transport theories.

1.3 *The regulation of the Alpine Rhine*

The Alpine Rhine upstream of the Lake of Constance forms the boarder between Switzerland and Austria. The regulation project envisaged at the end of 19th century proposed systematic narrowing and straightening, and in particular two cutoffs, one on Austrian, one on Swiss ground (see Meyer-Peter and Lichtenhahn, 1963). The Fussach cutoff which was realised between 1895 and 1900, shortened the distance to the lake and was thus supposed to enable the Rhine to deposit his entire sediment load there for a long time. The Diepodsau meander cutoff was realised in 1923. Instead of the expected bed lowering, rising of the bed level occurred. The unexpected performance of this cutoff contributed to the development of research on bed load transport in Switzerland.

2   THE MEYER-PETER/MUELLER BED LOAD FORMULA

2.1 *Meyer-Peter and hydraulic research*

Eugen Meyer-Peter was appointed as professor for hydraulic structures at the Swiss Federal Institute of Technology (ETH) in 1920. He had been working as an engineer mainly on port constructions. As a professor, he had to teach fluid mechanics, hydraulic structures, dam construction, soil mechanics and river engineering. According to Mueller (1953), Meyer-Peter found the theoretical basis of river engineering rather poor and felt the need for more basic research, namely on bed load transport.

Since he became a professor, Meyer-Peter urged for the building of a hydraulic laboratory. This vision was achieved in 1930 when the Hydraulic Laboratory of the Swiss Federal Institute of Technology was inaugurated. Immediately, basic research on bed load transport was undertaken. In parallel, the Laboratory was commissioned by the Federal Authority on Construction Inspection (Eidg. Oberbauinspektorat) in December 1931 to investigate the problems of the Diepoldsau cutoff, and the sediment transport problems of the Alpine Rhine in general. This contract was welcome, since the economic depression of that time slowed down other activities, namely hydropower development.

The new laboratory attracted a team of eminent scientists. In the first years of existence, therefore a lot of basic research on sediment transport was carried out, as well as specific investigation on the Alpine Rhine River.

2.2 *Hans Albert Einstein*

Born in Berne 1904 as son of the famous physicist Albert Einstein and Mileva Marič, Hans Albert Einstein grew up mainly in Zurich, where he visited high school. His class mates called him 'Steinli', what may be translated by 'pebble' (Roboz Einstein, 1991). This may be seen as an early sign of his later career. H.A. Einstein graduated at ETH as a civil engineer and worked a few years in Dortmund as a structural engineer. His father, of Jewish-German origin, was concerned about the political evolution in Germany. He therefore wrote a letter to Meyer-Peter, recommending his son for an employment. H.A. Einstein got involved in the research on bed load transport. He developed the wall drag correction procedure still named after him on the base of field velocity measurements (Einstein, 1934); and carried out tests for calibration of a bed load sampler (Einstein, 1937). Finally, he prepared a doctoral thesis on bed load transport, seen as a probability problem (Einstein, 1936). This work can be seen as the basis for his later work and the development of his own bed load transport formula (Einstein, 1950; Ettema and Mutuel, 2004). H.A. Einstein left Switzerland in 1938 for the United States. It was again his father who urged him to do so, because of the same political reasons.

2.3 Henry Favre, Erwin Hoeck and Robert Mueller

Henry Favre acted as vice director. He was mainly dealing with fluid mechanics, but was involved in the bed load transport research. Later, he became professor for general mechanics. Erwin Hoeck mainly carried out the model tests on the Alpine Rhine. Robert Mueller joined the team in 1931 and stayed at the laboratory until 1955. He was involved in river engineering and torrent control problems and participated at all the research work for bed load transport till the publication of the 1948-formula (see Hager, 2012).

2.4 The 1934-formula

On the basis of systematic tests Meyer-Peter et al. (1934) published a first bed load transport formula:

$$\frac{q^{2/3}J}{d} = 17 + 0.4\frac{q_B^{2/3}}{d} \tag{1}$$

where q is the discharge per meter channel width [m³/ms], q_B the bed load transport rate per unit width [kg/ms] and d the grain size [m]. At that time, little was known about bed load transport. Du Boys (1879) had proposed a relation based on shear stress. It was however based on a wrong concept of sediment transport movement and did not allow obtaining quantitative results. Gilbert (1914) had carried out bed load transport tests, but without proposing a transport equation. Schocklitsch (1934) had proposed such an equation just a little bit earlier than the Meyer-Peter group. In their publication, Meyer-Peter et al. derived first their equation from their own data set. They point out that the equation is compatible with the generalised Froude law of similitude. Then they compare their equation to the Gilbert data and find good agreement. Later, they compare their data to the Schocklitsch equation and find a systematic deviation between this formula and their data.

This research was accompanied by field measurements of sediment transport in the Alpine Rhine (Nesper, 1937). Mueller (1937) compared the results with the 1934-formula and found coincidence, at least for the order of magnitude.

2.5 The 1948-formula (the 'Swiss formula')

Over many years, bed load transport tests were continued. The biggest flume used was 2 m wide and 50 m long. Experiments were run for several days. Sediment feeding had mostly to be done by a person staying constantly on the flume. All this work finally led to the well-known 1948-formula (Meyer-Peter and Mueller, 1948). According to Yalin (1997), this formula is often called the 'Swiss formula', what fits well to the title of this publication.

Meyer-Peter and Mueller published their formula in the form (presented here in SI units):

$$\rho_W g\left(k_s/k_r\right)^{1.5}R_bJ = 0.047\rho_s g(\rho_s - \rho_w)d_m + 0.25\rho_w^{1/3}q_B^{2/3}g^{2/3}\left(\rho_s - \rho_w\right)^{2/3} \tag{2}$$

where ρ_w is the densitiy of water [kg/m³], g the gravity acceleration, k_s the Strickler coefficient of the bed [m$^{1/3}$/s], k_r a Strickler grain roughness coefficient [m$^{1/3}$/s] (calibrated on Nikuradse's 1933 tests), R_b the hydraulic radius of the partial bed section [m], determined according to Einstein's 1934 procedure, ρ_s the density of the bed material [kg/m³], and s the relative density of the solid to the water. d_m is the mean grain size, calculated as a weighted average from a grain size distribution curve. While the 1934-formula referred to a single grain size, the new formula applies also to grain mixtures.

Using the dimensionless parameters

$$\theta = \frac{R_b J}{(s-1)d_m} \tag{3}$$

and

$$\phi = \frac{q_b}{\rho_s \sqrt{(s-1)gd_m^3}} \tag{4}$$

the formula takes the well-known form

$$\phi = 8(\theta' - 0.047)^{1.5} \tag{5}$$

where

$$\theta' = (k_s/k_r)^{1.5}\theta \tag{6}$$

Note that recent publications show that in case of plane bed conditions the coefficient in equation (5) should rather be about 5 (Hunziker, 1995; Hunziker and Jaeggi, 2002; Wong and Parker, 2006).

2.6   Discussion

Prior to the work discussed above, there was no widely accepted bed load transport equation. Based on a large number of tests and respecting the laws of similitude between laboratory flumes and prototype, the 1948-formula has had a wide impact on river engineering and sediment transport research. Of course, many other equations have appeared since then; and may be more suitable than the Meyer-Peter/Mueller equation for certain conditions. But the work of Meyer-Peter and his team induced a wide acceptance of sediment transport relations based on laboratory tests and can thus be considered as a milestone in research on bed load transport.

3   SELF-ARMOURING

3.1   Development of armour layer

Meyer-Peter and his co-workers recognised self-armouring processes during their tests and in the analysis of the processes in the Alpine Rhine (Meyer-Peter and Lichtenhahn, 1963); but did not attempt to quantify their influence. Gessler (1965, 1970) performed self-armouring tests with constant slope and increasing discharge and monitored the change in surface grain size distribution. He then developed a procedure to define the grain size distribution of the armour layer in function of the original bed material distribution and the flow conditions. Guenter (1971) performed tests with rotational erosion. The bed stabilised at a certain slope with an armour layer at the surface. He proposed a procedure for the grain size distribution of the armour layer for these limiting conditions at the limit of stability.

Very often, it is sufficient to use d_{90} of the bed material (90% by weight of the material is finer than this diameter) as mean grain size of a fully developed armour layer.

3.2 Stability of armour layer

Gessler (1973) gave a criterion of stability for an armour layer, based on his procedure. Guenter (1971) proposed such a formula, which is very simple if d_{90} is taken as the mean grain size of the armour layer:

$$\theta_D = \theta_{cr} \left(\frac{d_{90}}{d_m} \right)^{2/3}$$
(7)

θ_{cr} is the Shields-factor for beginning of motion of uniform material, equal to 0.047 according to Meyer-Peter and Mueller for hydraulically rough conditions. Guenter's threshold value θ_D is the limit of erosion. In the absence of sediment supply, a bed may remain stable up to this value due to self armouring. Between the Shields and the Guenter threshold, there is no unique bed load function. Equations as (1) and (2) give in this range the upper limit of possible bed load transport and thus the limit to aggradation. Considering only bed-load transport formulas and neglecting the self-armouring process leads to an overestimation of erosion rates.

4 RESEARCH FROM 1980 TO 2000

4.1 Bed load transport on steep slopes

The tests run by Meyer-Peter and his co-workers had a maximum slope of 2%. The extension of transport formula to steeper slopes was a challenge the Laboratory of Hydraulics of ETH accepted in the eighties, fully in the Meyer-Peter tradition. On the basis of test runs in a steep flume with slopes up to 20%, and including the Meyer-Peter data, a bed load transport formula was proposed by Smart and Jaeggi (1983; see also Smart 1984):

$$\phi = 4 \cdot \left(\frac{d_{90}}{d_{30}} \right)^{0.2} J^{0.6} \frac{v_m}{v_*} \sqrt{\theta} \cdot (\theta - \theta_{cr})$$
(8)

v_m is the mean velocity of the flow, v_* the shear velocity and d_{30} the diameter corresponding to 30% of the bed material by weight.

Later, Rickenmann (1990) used the same flume to investigate the effect of fine sediment concentration in the flume. His investigations resulted in the formula

$$\phi = \frac{3.1}{\sqrt{s-1}} \left(\frac{d_{90}}{d_{30}} \right)^{0.2} \sqrt{\theta} \cdot (\theta - \theta_{cr}) \cdot Fr^{1.1}$$
(9)

Fr ist the Froude number. In this case, s is the relative density of the sediment to the slurry. It may vary between 1.2 and 2.65.

4.2 Bed load transport analysis with numerical models

The progress of computer technology allowed developing numerical simulation models to analyse bed load transport. The Laboratory of Hydraulics concentrated first on the sediment regime of the Emme river (VAW, 1987). Observed erosion seemed to indicate a recent instability due to unknown reasons.

Discharge is recorded and the bed level changes surveyed about every ten years. Using the quasi-steady and one-dimensional model MORMO, bed deformations could be calculated in function of the hydraulic conditions and sediment transport capacity. The only unknown parameter was the sediment input at the upstream boundary. This parameter was varied until the recorded bed level changes could be reproduced. In the Emme study, it could be shown that erosion was clearly induced by the river regulation and that bed load transport was strongly in disequilibrium, despite comparatively small annual bed level changes.

In a way, such models now allow to fully use the bed load formula like Meyer-Peter and Mueller's, developed decades earlier.

4.3 Grain sorting and fractionwise transport

Already Einstein (1950) recognised the different behaviour of grains in a mixture and included a hiding function in his procedure. Development of computer technology allowed investigating grain sorting and fractionwise transport numerically. Hunziker (1995) calibrated his procedures on the self armouring tests of Gessler (1965) and Guenter (1971), and the tests run by Suzuki and Hano (1992) where a mobile armour layer formed. Hunziker applied a modified version of the Meyer-Peter/Mueller formula to the mean grain size of the surface layer. For mobile armour layer conditions, Hunziker had to calibrate his hiding function respecting equal mobility of subsurface material and the moving bed load for mobile armour conditions. He used the Suzuki/Hano tests as well as the Guenter tests during rotational erosion before formation of a stable armour layer.

Using Hunziker's hiding function considerably allows improving the simulation of bed load transport in rivers, in particular for aggrading conditions.

4.4 Bed load transport in wide channels

When a bed load transport rate is calculated according to the above described formulas, for constant parameters but varying the channel width, the transport capacity reaches a maximum value for a certain width. For narrow channels, wall drag plays an important role and reduces transport capacity. For larger widths, the transport capacity decreases and may even reduce to zero. However, in reality a very wide river is usually braided. Partial channels are still able to move sediment. The total transport capacity of the river is then the sum of the transport capacity of the partial channels and becomes independent of the total channel width.

Zarn (1997) developed two procedures for such situations. The simpler one consists in determining first the function between transport capacity and channel width as described and the maximum value Q_{Bmax} for the given conditions. The transport rate for a very wide channel Q_B becomes

$$Q_B = Q_{B\,max}\left(3.65 e^{-0.86\frac{B_S}{B_m}} - 4\, e^{-1.5\cdot\frac{B_S}{B_m}} + 0.35\right)\frac{B_S}{B_m} \qquad (10)$$

B_s is the considered channel width and B_m the channel width corresponding to the maximum transport rate.

River restoration has become an important issue over the last years. A particular aim is often to widen the channel. It is then important to assess bed load transport in such wide channels.

5 RIVER SEDIMENT REGIME, THE CASE OF THE ALPINE RHINE

The problems encountered with the Alpine Rhine regulation and in particular the Diepoldsau cutoff had triggered the research work on bed load transport described. They remained a challenge for the researchers during the following decades.

After having developed the 1934 formula, Meyer-Peter recommended to narrow the main channel in the cutoff and to insert an inner dyke along its banks. It was recognised that these measures would generally increase bed shear stress and thus transport capacity. It was hoped that further aggradation could be prevented. The works were started in 1943 and accomplished by 1954.

On the basis of the 1948-formula, it was attempted to define an equilibrium longitudinal profile for the Alpine Rhine River. Meyer-Peter and Lichtenhahn (1963) summarized this work. Before the widespread use of computers, the authors used refined mathematical tools on one side, but had to make simplifying assumptions. Assuming a rather high abrasion coefficient, they found that an equilibrium profile could almost be fitted to the existing profile. Since the bed level was considered to be too high and flood capacity insufficient, dredging was envisaged.

Zarn et al. (1995) report on the numerical simulation of the bed load transport in the Alpine Rhine river. The model was calibrated on the observed bed level changes between 1974 and 1988. Like for the Emme river, the sediment supply from the tributaries and the upper reach could be assessed, by reproducing the observed bed deformations. In the sixties, gravel quantities dredged from the river exceeded those recommended by Meyer-Peter by far, so that general erosion occurred. In 1970, a road bridge collapsed and the dredging was then reduced or completely stopped. Conditions during calibration period were thus quite different from those encountered in the thirties or the fifties.

It was found that in the upper half erosion predominated and would also continue in the near future. Bed load supply from the upper river reach and the tributaries had also decreased, because of local dredging or hydropower schemes. In the lower half of the modelled reach aggradation dominated. Aggradation rates are however small and do not compensate the bed incision after the dredging for decades.

6 CONCLUSIONS

Research on bed load transport led by Eugen Meyer-Peter, Hans Albert Einstein, Robert Mueller and others between 1930 and 1948 in Zurich constitutes a milestone for a rational approach of this process. In the tradition of this work, further developments have been achieved, namely at the hydraulic laboratories of Federal Institutes of Technology in Zurich and Lausanne. The results presented here are just a selection. Research work on bed load transport is still going on in Switzerland.

REFERENCES

Du Boys, P., 1879, Etudes du régime du Rhône et l'action exercée par les eaux sur un lit à fond de graviers indéfiniment affouillable; *Annales des Ponts et Chaussées*, ser. 5, 18, pp. 141–195.

Einstein, H.A., 1934, Der hydraulische oder Profilradius, *Schweizerische Bauzeitung*, Bd. 103, Nr. 8, 89–91.

Einstein, H.A., 1936, *Der Geschiebetrieb als Wahrscheinlichkeitsproblem*, Promotionsarbeit, Eidgenössische Technische Hochschule Zürich, Druckerei Leemann & Co.

Einstein, H.A., 1937, Die Eichung des im Rhein verwendeten Geschiebefängers, *Schweizerische Bauzeitung*, Bd. 110, Nr. 12 bis 15, Sept./Okt., 29–32.

Einstein, H.A., 1950, The bed-load function for sediment transportation in open channel flows, *Tech. Bulletin No. 1026*, U.S. Dept. of Agriculture, Soil Conservation Service, Washington, D.C.

Ettema, R. and Mutuel, C., 2004, Hans-Albert Einstein: Innovation and compromise in formulating sediment transport by rivers, *Proc. ASCE, J. of Hydr. Engineering*, June, 477–487.

Gessler, J., 1965, Der Geschiebetriebbeginn bei Mischungen untersucht an natürlichen Abpflästerungserscheinungen in Kanälen; *Mitteilung der Versuchsanstalt für Wasser—und Erdbau*, No. 69, ETH Zürich.

Gessler, J., 1970, Self stabilizing tendencies of alluvial channels; *Proc. ASCE, J. of Waterways Div.*, Vol. 96, WW 2, April 1970, pp. 235–249.

Gessler, J.: 1973, Critical shear stress for sediment mixtures in rivers; *Int. Association for Hydraulic Research, Int. Symp. on River Mech.*, Bangkok.

Gilbert, G.K., 1914, Transportation of debris by running water, *Professional Paper No. 86*, U.S. Geological Survey, Washington, D.C.

Guenter, A.: Die kritische mittlere Sohlenschubspannung bei Geschiebemischungen unter Berücksichtigung der Deckschichtbildung und der turbulenzbedingten Sohlenschubspannungen; *Mitteilung der Versuchsanstalt für Wasserbau, Hydrologie und Glaziologie* Nr. 3, ETH Zürich 1971.

Hager, W., 2012, Eugen Meyer-Peter und die Versuchsanstalt für Wasserbau, *Wasser Energie Luft*, 104. Jahrgang, Heft 4, 307–313.

Hunziker, R.P., 1995, Fraktionsweiser Geschiebetransport, *Mitteilung der Versuchsanstalt für Wasserbau, Hydrologie und Glaziologie Nr. 118*, ETH Zürich.

Hunziker, R.P., and Jaeggi, M.N.R., 2002, Grain sorting processes, *Proc. ASCE, Journal of Hydraulic Engineering*, Vol. 128, Issue 12, 1060–1068.

Meyer-Peter, E., Favre, H., and Einstein, H.A, 1934, Neuere Versuchsresultate über den Geschiebetrieb, *Schweizerische Bauzeitung*, Bd. 103. Nr. 4, 89–91.

Meyer-Peter, E. and Mueller, R., 1948, Formulas for bed-load transport, *Proc. Int. Association for Hydraulic Research*, 2nd meeting, Stockholm, Sweden.

Meyer-Peter and Lichtenhahn, C., 1963, Altes und Neus aus dem Flussbau, *Eidg. Departement des Innern, Veröffentlichungen des Eidg. Amtes für Strassen—und Flussbau*, Eidg. Drucksachen—und Materialzentrale, Bern.

Mosonyi, E. (ed.), 1970, Johann Gottfried Tulla. *Ansprachen und Vorträge zur Gedenkfeier und Internationalen Fachtagung über Flussregulierungen aus Anlass des 200. Geburtstages.* Theodor-Rehbock-Flusslaboratorium, Karlsruhe.

Mueller, R., 1937, Überprüfung des Geschiebegesetzes und der Berechnungsmethode der Versuchsanstalt für Wasserbau an der ETH mit Hilfe der direkten Geschiebemessungen am Rhein, *Schweizerische Bauzeitung*, Bd. 110, Nr. 12 bis 15, Sept./Okt. 33–36.

Mueller, R., 1953, Flussbauliche Versuche an der Versuchsanstalt für Wasserbau an der ETH, in Eugen Meyer-Peter, Festschrift zu seinem 70. Geburtstag. *Vorausdruck der Schweizerischen Bauzeitung*, Zürich.

Nesper, F., 1937, Ergebnisse der Messungen über die Geschiebe—und Schlammführung des Rheins an der Brugger Rheinbrücke, *Schweizerische Bauzeitung*, Bd. 110, Nr. 12 bis 15, Sept./Okt.

Nikuradse, J.: 1933, Strömungsgesetze in rauhen Röhren; *Forschungsgebiet des Ing.-Wesens*, Heft 361.

Richenmann, D., 1990, Bedload transport capacity of slurry flows at steep slopes, *Mitteilung der Versuchsanstalt für Wasserbau, Hydrologie und Glaziologie*, ETH Zürich Nr. 103.

Roboz Einstein, E., 1991, *Hans Albert Einstein, Reminiscences of His Life and Our Life Together*, Iowa Institute of Hydraulic Research, The University of Iowa.

Schocklitsch, A., 1934, Geschiebetrieb und die Geschiebefracht, *Wasserkraft & Wasserwirtschaft*, Jgg. 39, Heft, 4.

Smart G.M., and Jaeggi, M., 1983, Sediment transport on steep slopes, *Mitteilung der Versuchsanstalt für Wasserbau, Hydrologie und Glaziologie*, ETH Zürich, Nr. 64.

Smart G.M., 1984, Sediment transport formula for steep channels, *Proc. ASCE, J. of Hydr. Eng.*, Vol 110, No. 3, 267–276.

Suzuki, K., and Hano, A. 1992, Grain size change of bed surface layer and sediment discharge on an equilibrium river bed, Proc. of the International Grain Sorting Seminar, *Mitteilung der Versuchsanstalt für Wasserbau, Hydrologie und Glaziologie*, ETH Zürich Nr. 117, 151–156.

Versuchsanstalt für Wasserbau, Hydrologie und Glaziologie der ETH Zürich und Geographisches Institut der Universität Bern, 1987, Emme 2050, *Studie über die Entwicklung des Klimas, der Bodenbedeckung, der Besiedlung, der Wasserwirtschaft und des Geschiebeaufkommens im Emmental, sowie über die Sohlenentwicklung und den Geschiebehaushalt in der Emme und zukünftige Verbauungskonzepte*, im Auftrag der Baudirektion des Kantons Bern und des Baudepartements des Kantons Solothurn (unpublished report).

Vischer, D., 1986, Schweizerische Flusskorrektionen im 18. Und 19. Jahrhundert, *Mitteilung der Versuchsanstalt für Wasserbau, Hydrologie und Glaziologie*, ETH Zürich, Nr. 84.

Wong, M., and Parker, G., 2006, Reanalysis and correction of bed-load relation of Meyer-Peter and Müller using their own database, *Proc. ASCE, Journal of Hydraulic Engineering*, Vol. 1132, No. 11, November, 1159–1168.

Yalin, M.S. 1977. *Mechanics of Sediment Transport*, 2nd edition, Oxford, Pergamon Press.

Zarn, B., 1997, Einfluss der Flussbettbreite auf die Wechselwirkung zwischen Abfluss, Morphologie und Geschiebetransportkapazität, *Mitteilung der Versuchsanstalt für Wasserbau, Hydrologie und Glaziologie*, ETH Zürich, Nr. 154.

Zarn, B., Oplatka, M., Pellandini, St., Mikoš, M., Hunziker, R., Jaeggi, M., 1995, Geschiebehaushalt Alpenrhein, Neue Erkenntnisse und Prognosen über die Sohlenveränderungen und den Geschiebetransport, *Mitteilung der Versuchsanstalt für Wasserbau, Hydrologie und Glaziologie*, ETH Zürich, Nr. 139.

Swiss Competences in River Engineering and Restoration – Schleiss, Speerli & Pfammatter (Eds)
© 2014 Taylor & Francis Group, London, ISBN 978-1-138-02676-6

Freeboard analysis in river engineering and flood mapping—new recommendations

L. Hunzinger

Flussbau AG SAH, Bern, Switzerland (On behalf of the Commission for Flood Protection of the Swiss Association for Water Management)

ABSTRACT: The Swiss Commission for Flood Protection elaborated a unified concept to determine the freeboard in order to evaluate the discharge capacity of a stream. It consists of several partial freeboards. On one hand, the partial freeboards take the uncertainty of the calculation of the water level into consideration. On the other hand, they consider hydraulic processes like the formation of waves, the backwater effect at obstructions or the additional space needed to convey floating debris underneath bridges. The concept includes recommendations on how to account for freeboard in flood mapping. With its concept paper on freeboard, the Swiss Commission for Flood Protection wishes that the freeboard will be taken into account in flood mapping and hydraulic design in Switzerland on an unified and coherent basis.

1 INTRODUCTION

Design procedures for flood protection structures or the evaluation of the discharge capacity of a stream usually consider a freeboard. However, in Switzerland no best practice has been established so far. Some practitioners use a constant value (e.g. 1 m) others set the required freeboard equal to the velocity head of the current. Furthermore, different criteria and different approaches are being used in design procedures and hazard evaluations (KOHS, 2012). This multitude of approaches seeds doubts among design engineers and authorities about the use of the "correct" freeboard and makes it difficult to compare different hazard analysis and flood protection projects. In order to overcome this weakness, the Swiss Commission for Flood Protection elaborated a unified concept for the determination of the freeboard. The concept is applicable to river courses. The freeboard requirements for dams and reservoirs in Switzerland are defined in the respective guidelines (BWG, 2001). The present publication is an extract of the recommendation published by KOHS (2013a) in German and KOHS (2013b) in French.

2 THE FREEBOARD CONCEPT

2.1 *Definitions*

The freeboard f denotes the vertical distance between the water level and the top edge of the bank or a hydraulic structure (Fig. 1) or the bottom edge of a bridge (Fig. 2). The water level can be observed or calculated. The required freeboard f_r denotes the freeboard that is necessary to guarantee a calculated discharge capacity.

2.2 *Uncertainty, wave formation and back water effects*

On one hand, the required freeboard is understood as a parameter that describes the uncertainty in the calculation of the water level for a given cross section geometry. On the other

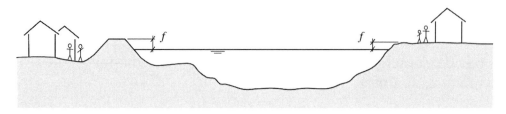

Figure 1. The freeboard f denotes the vertical distance between the water level and the top edge of the bank.

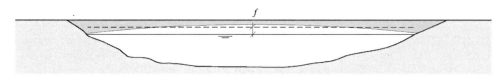

Figure 2. The freeboard f denotes the vertical clearance between the water level and the mean bottom edge of a bridge.

hand it considers wave formation or local backwater effects that are not necessarily included in hydraulic calculations. The required freeboard is therefore conceived as a *hydraulic* parameter. It should not be used to cover uncertainties in hydrologic peak flood estimation nor to justify an elevated flood protection objective for high damage potentials.

The required freeboard covers uncertainties in the calculation of the water level that have their origin in the uncertainties of

- The measured cross section geometry,
- The calculated bed level during peak discharges,
- The determination of the channel roughness and
- The determination of the effective channel geometry in presence of growing vegetation.

The above-mentioned uncertainties must be displayed as a result of the hydraulic calculations. They should not be replaced by applying conservative values for channel roughness or the channel geometry.

The required freeboard is used to cover the following processes and it ensures that the discharge capacity is not exceeded despite of these phenomena:

- Waves that are formed by the current (namely at flow conditions near to critical flow).
- Drift wood and drift ice.
- Back water effects at local obstacles (e.g. trees or overhanging corners of walls).

Sediment deposits at the channel bottom during floods, the banking of the water level in bends or the accumulation of drift wood and drift ice at bridge piers and abutments raise the water level. These effects must be considered when calculating the water level. They may not be regarded as effects covered by the freeboard.

2.3 *Load, impact and capacity*

In a given river course, the discharge Q and the supply of bed load S_b, drift wood and other floating debris F_d may be considered as loads (Fig. 3). As impact parameters the water level z_w, the flow velocity U and a parameter d above the water level can be defined. The latter describes the space occupied by floating debris. The water level is a result of the (changing) bed level z_b and the flow depth h. The cross section area A, the channel slope S, its roughness k and the freeboard f determine the capacity of the cross section to convey the load parameters.

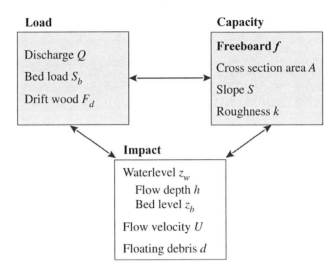

Figure 3. The system of load, impact and capacity sets the framework of the freeboard concept.

3 CALCULATION OF THE REQUIRED FREEBOARD

3.1 *Partial freeboards*

The required freeboard consists of three partial freeboards that are added geometrically. Each of the partial freeboards takes into account one of the above-mentioned impacts parameters.

$$f_{min} \leq f_r = \sqrt{f_w^2 + f_U^2 + f_d^2} \leq f_{max} \qquad (1)$$

where f_{min} = minimal required freeboard, f_{max} = maximum required freeboard, f_w = required freeboard due to uncertainties in the calculation of the water level, f_U = required freeboard due to wave formation and back water effects caused by local obstacles and f_d = required freeboard due to additional space needed to convey drifting debris underneath bridges.

3.2 *Freeboard due to uncertainties in the calculation of the water level*

The partial freeboard f_w is set equal to the uncertainty of the water level calculation σ_w

$$f_w = \sigma_w = \sqrt{\sigma_{wb}^2 + \sigma_{wh}^2} \qquad (2)$$

This uncertainty has two reasons (Fig. 4): first, the estimated bed level z_b during peak discharges may have an error and this error affects the water level calculation (σ_{wb}), second, the calculation of the flow depth h above the bed level may be imprecise because the cross section geometry may not represent the channel geometry properly or the roughness coefficients may be badly estimated (σ_{wh}). Both errors are added geometrically.

In Figure 4 σ_{wb} marks the error of the bed level estimation. It is set equal to its contribution to the water level calculation. The value of σ_{wb} must be estimated case by case. Values between 0.1 m in large low land rivers and 1.0 m in torrents are typical.

In order to determine the uncertainties σ_{wh}, flow depth was calculated in 18 rivers in Switzerland using the Manning-Strickler-formula and assuming errors of the input parameters as follows: channel width ±10% (max. ±1 m); measured bed level ±0.1 m, roughness coefficient ±10%, longitudinal slope ±10%, bank slope ±3°. The errors of the independent input parameters were propagated to the dependent variable flow depth. In total 52 flow depths at different discharges were calculated. In Figure 5 a strong dependency of the uncertainty from the flow

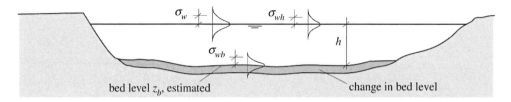

bed level z_b, estimated   change in bed level

Figure 4. Uncertainty in water level calculation. The flow depth h is calculated on an estimated bed level z_b, that varies with discharge and time. Both the estimation of the bed level and the flow depth calculation have errors.

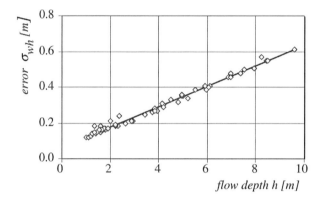

Figure 5. Error σ_{wh} of the flow depth calculation depending on the flow depth h.

depth itself can be observed. The error σ_{wh} of the water level calculation caused by uncertainties of the flow depth calculation can therefore be estimated using the following equation:

$$\sigma_{wh} = 0.06 + 0.06\,h \qquad (3)$$

Given certain circumstances it might be worth calculating σ_{wh} case by case using an estimation of the errors of the input parameters instead of applying equation 3.

3.3 Freeboard due to wave formation and backwater effects

Obstacles in the cross section (e.g. bridge piers, abutments, overhanging corners of walls) have a local backwater effect. The water level may rise to the level of the energy line. It is therefore in maximum $U^2/2g$ above the water level.

At flow conditions close to critical, waves appear where flow velocity is at maximum. That is in straight channels in the middle of the cross section. The wave crest lies at most to the extent of $U^2/2g$ above the mean water level.

The partial freeboard due to wave formation and backwater effects is therefore given by

$$f_U = \frac{U^2}{2g} \qquad (4)$$

where U = flow velocity and g = acceleration due to gravity.

3.4 Freeboard due to additional space needed underneath bridges

Flow underneath a bridge needs additional space to convey floating debris (drift wood, drift ice etc.) without clogging. In order to determine the partial freeboard f_d for wooden debris

34

Table 1. Partial freeboard f_d for wooden debris.

	f_d at bridges with a smooth bottom view (m)	f_d at bridges with a rough bottom view (m)
Small wooden debris (branches only)	0.3	0.5
Tree trunks, drifting individually	0.5	1.0
Rootstocks	1.0	1.0
Tree trunks, drifting as a carpet	1.0	1.0

Table 2. Criteria to apply the partial freeboards.

Partial freeboard	Criteria
f_w	In all river reaches.
f_U	At bridge cross sections; In reaches with flood protection dikes or walls that may collapse as they are overtopped; In reaches where slopping the banks results in a considerable water outlet; On alluvial fans; In paved torrent trenches.
f_d	At bridge cross sections where floating debris are relevant.

a system of classes is proposed. f_d has a value between 0.3 m and 1.0 m and depends on the characteristics of the drift wood and on the construction type of the bridge (Table 1). In rivers with other floating debris than wood (e.g. drift ice), f_d must be defined accordingly.

3.5 Selection of partial freeboards

The calculation of the required freeboard can be adapted to the river reach of interest by selecting the relevant partial freeboards f_w, f_U and f_d. According to the given situation one or two of the partial freeboards can be set to zero. The criteria given in Table 2 should be applied.

3.6 Minimum and maximum required freeboard

The required freeboard should be calculated cross-section-by-cross-section and should be unified along river reaches. A minimum value of the required freeboard of 0.3 m should be used. This gives more weight to the uncertainty of the calculated water level in small, slowly flowing rivers. A maximum value of the required freeboard prevents unrealistic high values. In watercourses with fluvial bed load transport, a maximum value of 1.5 m is proposed. In torrents with potential debris flow, the maximum value could be higher.

4 EFFECTS OF THE EXCESS OF CONVEYANCE

4.1 Freeboard and weak point analysis

The freeboard is used to determine weak points in the framework of the design of flood protection measures or in the framework of a hazard analysis. The weak point analysis gives answers to the following questions:

Where does water overtop the banks at a given discharge?
Why does overbank flow occur?
Which amount of water overflows?

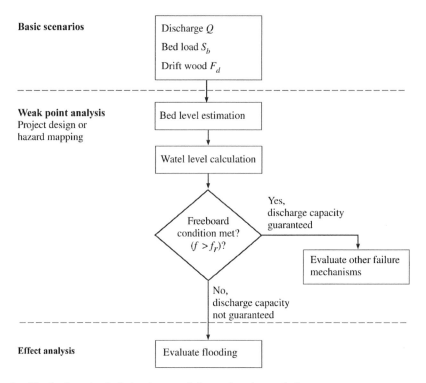

Figure 6.   The freeboard calculation is part of the weak point analysis.

The required freeboard f_r in a river reach is determined for a given discharge Q, bed load S_b and load of floating debris F_d. It does not matter whether the discharge Q corresponds to the design discharge of a hydraulic structure or the overload discharge of a flood protection measure or whether it corresponds to a flood scenario with a given return period.

If the calculated water level z_w (calculated with discharge Q) results in a freeboard f that is larger than the required freeboard f_r, the capacity of the river reach is sufficient to convey the loads without overbank flow. Using the terminology of hazard assessments the considered river reach is no weak point. No flooding is expected. Other types of failures must eventually be considered.

If the calculated water level z_w (calculated with discharge Q) results in a freeboard f that is lower than the required freeboard f_r, the capacity of the river reach is insufficient to convey the discharge loads without overbank flow. Using the terminology of hazard assessments the river reach in consideration is a weak point and flooding may occur.

4.2   Effects in river courses with overflow resistant banks

If the discharge capacity is considered insufficient in a river course that is delimited by naturally grown terrain or by a dike or wall that remains stable even if it is overtopped, an overbank flow scenario as a function of the water level can be defined. The relevant water level z_w' corresponds to the calculated water level including its error (Equation 5):

$$z_w' = z_w + \sigma_w = z_w + \sqrt{\sigma_{wb}^2 + \sigma_{wh}^2} \tag{5}$$

This approach allows defining a flooding scenario whenever the discharge capacity is considered insufficient (Fig. 7, to the right). An alternative approach that is often used in flood mapping using 2d-simulations defines the calculated (best estimate) water level z_w as the relevant water level. However, as shown in the example of Figure 8 this approach would

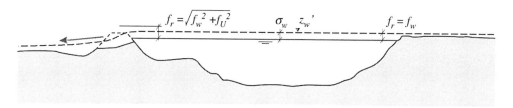

Figure 7.   Flooding scenario in case of excess of discharge capacity ($f < f_r$). To the left: dike break. To the right: flooding in case of a bank resistant to overflow. The required freeboard f_r along the dike (left) differs from that along the naturally grown bank (right) according to Table 2.

neglect overbank flow although the discharge capacity is declared insufficient because the freeboard condition is not fulfilled.

4.3   *Effects in river courses with banks not resistant to overflow*

If the discharge capacity is considered insufficient in a river course that is delimited by a dike or a wall that does not resist overtopping, a failure scenario for the dike or wall (Fig. 7 to the left) is defined. In order to calculate the outflow the water level, z_w' according to equation 5 or the best estimate of the water level z_w may be used. Usually, the outflow depends rather on the size of the dike breech, at the time of collapse or on sediment deposit in the channel than on the selection of the relevant water level.

5   CONCLUSIONS

The presented method for the determination of the required freeboard has to be considered as a recommendation by the Commission for Flood Protection of the Swiss Association for Water Management. It addresses hydraulic engineers of the private sector and of authorities. The approaches have been developed in an effort to be transparent, coherent and generally applicable. Nevertheless, the engineer is encouraged to adapt the method to the specific conditions of the river under consideration and hence to improve the approach.

The freeboard calculation is one element of the assessment of flood hazards or the design of hydraulic structures. Other elements like the definition of flood protection objectives, the definition of design scenarios or the concept to deal with overload scenarios must be examined separately.

ACKNOWLEDGEMENTS

The following members of the Commission for Flood Protection elaborated the recommendation on the freeboard: Lukas Hunzinger, Martin Jäggi, Jean-Pierre Jordan, Jürg Speerli, Heinz Weiss and Benno Zarn. The Swiss Office for Environment funded the elaboration process.

REFERENCES

BWG, 2001: Sicherheit der Stauanlagen. Richtlinien des Bundesamtes für Wasser und Geologie.
KOHS, 2012: Literaturrecherche Freibord. HSR Hochschule für Technik Rapperswil—*Institut für Bau und Umwelt*.
KOHS, 2012a: Freibord bei Hochwasserschutzprojekt und Gefahrenbeurteilungen. Empfehlung der Kommission für Hochwasserschutz (KOHS). *Wasser Energie Luft* 105(1): 43–50.
KOHS, 2012b: La revanche dans les projets de protection contre les crues et de l'analyse de dangers. Recommendations de la Commission pour la protection contre les crues (CIPC). *Wasser Energie Luft* 105(2): 122–129.

Swiss Competences in River Engineering and Restoration – Schleiss, Speerli & Pfammatter (Eds)
© 2014 Taylor & Francis Group, London, ISBN 978-1-138-02676-6

Alpine Rhine Project (Section River Ill—Lake Constance)

Markus Mähr
International Rhine Regulation IRR, St. Margrethen, Switzerland

Dominik Schenk, Markus Schatzmann & André Meng
Basler & Hofmann AG, Consulting Engineers, Esslingen, Switzerland

ABSTRACT:

Risks: The 26 km long international stretch of the Alpine Rhine River between the confluence of the River Ill and the estuary into Lake Constance was canalised at the end of the 19th as well as at the beginning of the 20th century in order to alleviate the problems caused by severe major floods in the preceding years. The river alterations enabled an economic as well as an agricultural boom in the lower Rhine River valley. Today approx. 300'000 people live and work within this region due to the rectification work by the International Rhine Regulation IRR. Within the last decades the geotechnical as well as the safety standards have changed along with ecological restoration issues. Due to the significant increase of urban areas, the potential risk and vulnerability have increased tremendously.

Flood control as a main goal: The main objective for the Alpine Rhine Project is an improvement of the current flood protection system. The design discharge of the project is 4'300 m³/s (equivalent to a 300-year flood), while also taking into account an extreme event of 5'800 m³/s. In order to achieve this goal, heavy construction measures with new levees and widening of the existing riverbed are necessary. All measures must fulfil today's requirements of both neighbouring countries Austria and Switzerland. This includes an improvement of the ecological situation, the preservation of the existing drinking water supply and a minimal impact on the environment during construction.

Ecology and flood protection—a double win: According to the European Water Framework Directive (WFD) the lower stretch of the Alpine Rhine River is considered to be "heavily modified". In order to fulfil the WFD the river needs to be ecologically restored. According to the Swiss Water Protection Act the natural river state must be achieved as far as possible. However, all parties involved agree that the original historic morphology of the Rhine River can never be fully restored. Therefore the main goal of the project is to allow as much lateral river widening as feasible. A wider cross section of the Alpine Rhine River allows both ecological restoration and the required flood protection level.

1 BACKGROUND

1.1 *History*

The original Alpine Rhine has often rebuilt its bed. It was free to meander in the valley plains without any constraints. Mud and bed load have been deposited, lakes, swamps and new gravel banks arose, old gravel banks were relocated. Floods due to sedimentation of the river bed were very common and are mentioned in the eleventh century for the first time.

The flood disaster with the biggest impact on the Rhine valley happened in 1817, when a large part of the Rhine estuary and wide areas of the valley were flooded. Further devastating floods occurred in 1888 and 1927. Due to the increasing urbanization of the Rhine valley, better flood protection was demanded in the nineteenth century.

The two border countries Austria and Switzerland started negotiations for a joint regulation in 1822. More than half a century and a series of heavy floods later, the first state treaty was signed, founding the "International Rhine Regulation". The main objectives of this treaty were:

- Two meander cut offs that reduced the stretch from 36 km to 26 km,
- New embankment dams and a new channel with a uniform width of 110 m.

Because of the increasing sediment aggradation in the Lake Constance bays of Fussach, Hard and Bregenz, a second treaty was signed in 1924. This second treaty established not only the continuation of the regulation works, but also the jetty of the regulation structures on the alluvial fan in Lake Constance in order to prevent further deposition. In 1954, a third treaty was necessary, as the discharge capacity was increased to 3'100 m³/s and the main channel was narrowed to approx. 70 m to reduce the costs of sediment management. The development of the Alpine Rhine River is also shown in Figure 1.

1.2 Situation today

Since the signing of the last treaty in 1954, the population living along the Rhine has more than doubled. Today, approx. 300'000 people live and work within this region due to the rectification work by the International Rhine Regulation (IRR). As a result, the accumulated potential damage in the project stretch increased to 5.7 Mrd. Swiss francs in 2008 (Niederer St., 2008).

According to the European Water Framework Directive (WFD), the lower stretch of the Alpine Rhine River is considered to be "heavily modified", as the Rhine shows deficit regarding quantity and diversity of aquatic life. For example only 6 out of 31 fish species are still numerously present, 11 species are endangered and 14 species are extinct. This is a result of the rectification and the water level fluctuations due to hydro peaking. In order to fulfil the

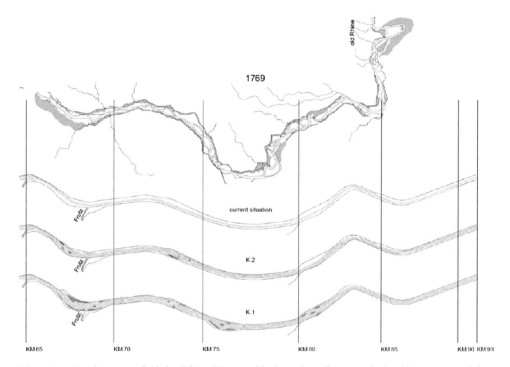

Figure 1. Development of Alpine Rhine River and its lateral confinement during history, natural situation 1769, current situation and situation with the first and second combined concepts K.1 and K.2. Picture: consortium "Zukunft Alpenrhein"/Flussbau AG.

WFD, the river needs to be ecologically restored. According to the Swiss Water Protection Act, the natural state must be achieved as far as possible. Due to the settlements and infrastructures, the original historic morphology of the Rhine River can never be fully restored. Thus, the main goal of the project is to allow as much lateral river widening as feasible. An increased cross section allows both ecological restoration and flood control.

Today the floodplains are maintained and cultivated by local farmers. A total of 450 ha on both river sides are rented out to farmers. Due to its flatness, fertility and accessibility, the land is well suited for grassland management.

The main aquifers in the Rhine valley are in the vicinity of the river, where the gravel layers are situated. Hence most of the wells are located along the dams or in the floodplains: 13 wells in Switzerland and 10 wells in Austria, providing the drinking water supply in the valley. The wells are surrounded by protection areas, whereof some extend even to the middle of the existing channel. Beside the utilizations outlined above, the floodplain is further used for leisure activities such as cycling, football, hiking, horse riding and dog sports. Figures 2 and 4 give an impression of the current situation and Figure 3 shows how the Alpine Rhine could look like in the future.

Figure 2. Current situation of Alpine Rhine River between Km 67 and Km 72 (view downstream). Rural Setting: Left: Agricultural area and village of Oberriet/Switzerland. Right: Rivermouth of river Frutz and village of Koblach/Austria (further down).

Figure 3. Visualization of combined concept 1 of Alpine Rhine River between Km 67 and 72. Maximum possible river widening in this river section: 300 to 500 m. Visualization: consortium "Zukunft Alpenrhein"/Beitl GmbH.

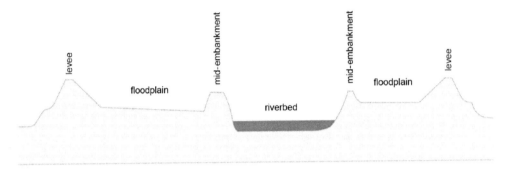

Figure 4.   Characteristic current profile of Alpine Rhine River in the project section.

2   PROJECT

2.1   *Project goals*

Based on the development concept of the alpine Rhine (Michor K., 2005) that revealed the shortcomings regarding flood protection as well as the ecological situation, the joint committee of the Rhine decided to initiate a flood protection project. The project goals were defined according to the recommendations of the development concept as follows:

- Increase flood protection up to (at least) 4'300 m^3/s,
- Maintain drinking water supplies,
- Improvement of the ecological situation,
- Economical use of resources.

2.2   *Project organization*

The project is promoted by the International Rhine Regulation (IRR) which is by definition submitted to international law. Given this situation, the following issues arise:

- The project has to fulfill the laws of both countries,
- The planning work will be financed fifty-fifty by the two countries,
- A new treaty has to be signed to establish future collaboration and funding of the project.

After a two stage tendering process under WTO guidelines, the planning consortium "Zukunft Alpenrhein" was appointed as preferred bidder in 2011. They started their work with the study of alternative concepts, based on the results of the feasibility study conducted by the Federal Institute of Technology Zurich in 2011.

To ensure acceptance by the public, this planning phase is accompanied by a partici-patory process. For that purpose, a communication committee manages public relations as well as participation. Six experts of various topics accompany the planning process. They act both as control of the planning consortium as well as consultants of the project management.

2.3   *Elaborating the project—a multidisciplinary teamwork*

The present project is elaborated by a consortium of eleven engineering and planning compa-nies from Switzerland and Austria. The team must include all disciplines, the local knowledge and provide the necessary engineering and designing capacity for a large scale river project. The involved companies in the consortium "Zukunft Alpenrhein" are: Basler & Hofmann AG, Geoconsult, Flussbau AG, Simultec, 3P Geotechnik, Beitl GmbH, AquaPlus, Oeplan GmbH, Hydra, Baenziger Partner AG, M&G Engineers.

3 CONCEPT PHASE

3.1 Design flood, extreme event and freeboard

The selected design flood has a return period of 300 years and a corresponding peak discharge of $Q = 4'300$ m³/s.

The extreme event of the project is a flood with a hypothetical recurrence period of 500 to 800 years and a corresponding peak discharge of $Q = 5'800$ m³/s.

While the design flood is considered for the planning of mitigation measures in order to protect the entire valley along the river section against damages and fatalities, the extreme event is used for the planning of further measures to prevent an entire collapse of the flood protection system and consequently minimize fatalities and damages.

The freeboard adopted for the design flood is calculated for the different river sections according to the recommendations of the Swiss commission for flood protection (KOHS) and according to the Austrian guideline. Both ways of calculating the freeboard take into account the uncertainties of the stream bed and the water level elevation, the energy head and for bridges the additional height due to floating wood. The average freeboard for the design flood is approximately 1.0 m. The freeboard for the extreme event is discussed in section 4 "handling the extreme event".

3.2 Investigated concepts

In a first phase, six basic concepts B.1 to B.6 were investigated, which resulted in two combined concepts K.1 and K.2 (Table 1) by focusing on the maximum possible morphological/ecological restoration (K.1) and by focusing on a first political optimum (K.2). After the evaluation of these two combined concepts, a third combined concept was derived as a technical, economical and political optimum (see section 3.5).

3.3 Hydraulic and morphological investigations

For each concept, extensive morphological and hydraulic investigations were carried out with the help of numerical simulations using the combined water-sediment transport model MORMO. The morphological model had been calibrated for a period of 10 years (1995–2005)

Table 1. Investigated basic and combined concepts.

Basic concepts:	
B.1	Elevation of main levees (+1.75 m)
B.2	Elevation of mid-embankments (+2.50 m), lowering of floodplain (down to 3.00 m above riverbed, elevation of main levees (+1.75 m)
B.3	Widening of riverbed from 70 m to 120 m
B.4	Widening of riverbed from 70 m to 170 m (in one section only up to 140 m because of limited river width), removal of mid-embankments
B.5	Maximum allowable widening of riverbed 10 m to the main levees (new riverbed width varies between 140 m and 370 m depending on the river section)
B.6	Lowering of riverbed (between 0 and 2.00 m), widening of riverbed from 70 m to 110 m
Combined concepts:	
K.1	Lateral shifting of main levees in selected river sections, preservation of main wells in the floodplain, maximum allowable widening of riverbed 10 m to the main levees (new riverbed width varies between 100 m and 500 m)
K.2	No lateral shifting of main levees, preservation of most wells along river section, widening of riverbed in selected river sections (new riverbed varies between 100 m and 300 m), preservation of more agricultural land in the floodplain
K.3	Technical, economical and political optimum (see section 3.5)

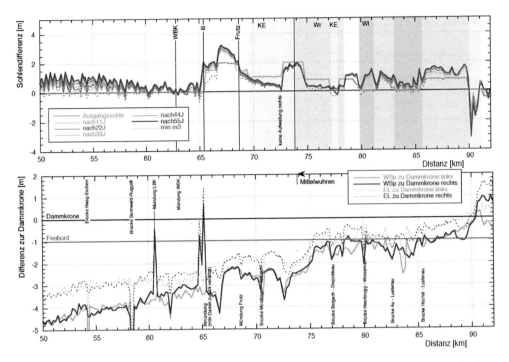

Figure 5. Combined Concept 2, Subconcept 2b (example). Longitudinal evolution of the riverbed over a period of 55 years (top) and water level and energy head compared to height of levee crest and freeboard (bottom). Optimized sediment excavation sites and yield at Km 70.6–71.0, Km 77.2–77.8 and Km 90.0. Investigations and pictures by consortium "Zukunft Alpenrhein"/Flussbau AG.

and validated for a period of 6 years (2005–2011). The hydraulic model had been calibrated based on the 23-8-2005-flood with a maximum discharge of 2'260 m³/s.

In a first step, the evolution of the riverbed was simulated for a future period of 55 years while considering today's sediment (gravel, sand) input (80'000 m³/year) and output (dredging, input into Lake Constance) by distinguishing increments of 11 years. In a second step, stationary hydraulic simulations of the design flood and the extreme event were carried out considering the riverbed after 55 years followed by the analysis of the water level and energy head in the longitudinal section (Fig. 5).

Some concepts, especially the combined concepts, were further investigated by distinguishing several subconcepts. The aim was to (1) optimize the sediment household and (2) minimize the elevation of the main levees by variation of the sediment dredging locations and the excavated volumes (m³/year).

With an annual input of 1.2 Mio m³ of fine sediment and an annual removal of 150'000 to 200'000 m³ on the floodplains, the study of fine sediment transport is relevant and was analyzed using the MORMO XO3 – model. It was shown that the river widening generally has a very positive effect on the reduction of fine sediment deposition and consequently river maintenance. However, the wider the river the higher the maximum scour depths. This in turn increases the costs for scour protection measures at the banks.

3.4 Rating and evaluation of concepts

After the hydromorphological investigations were completed, all other aspects of a given concept were analysed in the following fields of interest: drinking water/groundwater, geotechnics, water ecology and fishes in particular, terrestrial ecology, agriculture, recreation & leisure, etc. Considering all the fields of interest, a total of 25 quantitatively rateable criteria were defined while many of them include different subcriteria (Table 2). The 25 criteria, their

Table 2. Overview of evaluation criteria.

Field	Subject
Flood protection	Water: Degree of protection, behavior/consequences during extreme event, adaptability of measures
	Sediment transport and floating wood
Land use and management	Settlements and economic space
	Leisure and recreation
	Soil and agriculture
	Energy use
	Water supply
Environment	Morphology
	Aquatic and terrestrial ecology
	Water quality and regime
Realizability	Technical, legal and time-risks

specific targets and rating systems were elaborated in a long and intense process together with the team of experts.

Simultaneously, the costs were determined for each concept, analysing the main earth movements, river protection measures, dam rehabilitation measures, infrastructure costs (wells, bridges, roads, lifelines, etc.) and agricultural re-compensation, etc.

3.5 *Assessment of a technical, economical and political optimum—the third combined concept*

A draft of a third combined and optimized concept has been elaborated. Several investigations need to be carried out before any decision on this concept can be made and the concept can be finalized. Figure 6 gives an overview over priorities of different interests along the Rhine River.

Geotechnical investigations: A comprehensive investigation was carried out to study the levee foundation and the state of the existing levee while installing further measuring systems to determine the ground water level. Whereas the levee foundation proved to be mostly uncritical (except some well known sections of underlaying turf), the state of the levee proved to be more critical than assumed. As a result, an overview map was established which showed the levee state in three classes: Sections where immediate measures such as seal/slurry walls are required, sections where emergency measures such as filter loads shall be prepared for a flood event and sections where standard maintenance and monitoring is sufficient for the next few years until the Rhine River project will be built. In the final project, the levees need to be adapted on both river sides and on their entire lengths in order to meet today's safety standards.

Water supply: In a specific study, all wells were investigated in terms of importance within the water supply system of the Alpine Rhine valley. In the draft of the third concept, none of the wells must be given up except temporarily during river reconstruction. Nevertheless, a relocation of selected wells in the most narrow river sections would be beneficial for ecological and economic reasons (river width and levees height).

Agricultural planning: A larger area of grassland on the floodplains is going to be lost due to the River Rhine widening. An extensive study shall answer how this loss can be partly compensated with e.g. an amelioration of the agricultural turf land outside of the levees by ploughing excavated silt and sand material into the existing soil.

Further planning: The results of the above studies influence the final version of the third combined concept. The concerns of the communities and different NGO's must be considered. As several particular interests are contradictory, a long and intense technical and political debate will be necessary to achieve an optimal solution. It is a must that all involved parties and individuals are willing to abandon maximal claim in order to reach a consensus and the envisaged flood protection as soon as possible.

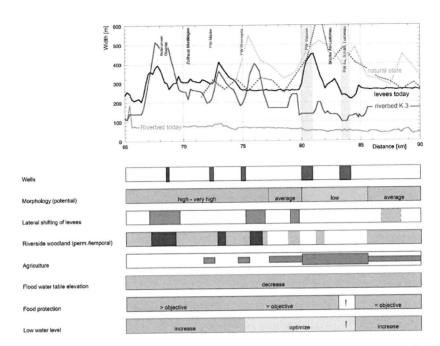

Figure 6.   Priorities of different interests along the Alpine Rhine River (combined concept 3). Picture: consortium "Zukunft Alpenrhein"/Flussbau AG.

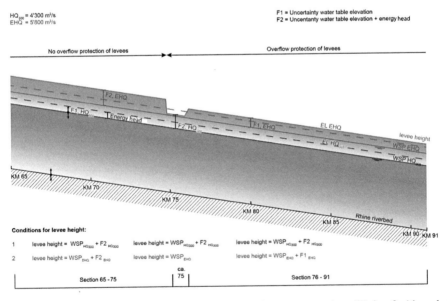

Figure 7.   Concept how to handle the extreme event. Picture: consortium "Zukunft Alpenrhein"/ Basler & Hofmann.

4   HANDLING THE EXTREME EVENT

In stationary 2D-simulations showed that the flooding of an extreme event would have a very large impact on the lower Alpine Rhine valley regarding the number of endangered people as well as the damage on infrastructure, industries and settlements. Several options were elaborated to minimize the impact of an event that exceeds the design flood. The following option proved to be the most robust and most cost-effective (Fig. 7): A) Km 65 to Km 74: New levees but no additional measures as the levees are high enough to provide the full freeboard recommended by KOHS even in the extreme event. B) Around Km 75 (rural setting): Levee on the same level as the water level of the extreme event. Spillway dam (overflow protection of dam) in order to avoid dam erosion followed by dam break. C) Km 75 to Km 91 (urban setting). Levee height to provide half of the freeboard as recommended by KOHS even in the extreme event. Overflow protection of dam in order to avoid dam erosion followed by dam break.

5 CONCLUSION AND OUTLOOK

The overall objective of the Alpine Rhine Project is to increase the flood protection level for the lower Alpine Rhine Valley. The aim of the concept phase is to find a technically, economically and politically optimal solution. The best concept shall be accepted by all parties involved, taking into account their—sometimes even contradictory—interests. All parties must therefore be willing to make compromises in order to reach a consensus and provide adequate flood protection as soon as possible. According to the current schedule, the best concept should be evaluated until summer 2015.

REFERENCES

Michor K. et. al. 2005. *Entwicklungskonzept Alpenrhein.*
Niederer St. und Pozzi A. 2008. *Schadenrisiken und Schutzmassnahmen im Alpenrheintal, Ergänzungsbericht A2+.*
Schweizerischer Wasserwirtschaftsverband. 2013. Freibord bei Hochwasserschutzprojekten und Gefahrenbeurteilungen, Empfehlungen der Kommission Hochwasserschutz (KOHS).

Swiss Competences in River Engineering and Restoration – Schleiss, Speerli & Pfammatter (Eds)
© 2014 Taylor & Francis Group, London, ISBN 978-1-138-02676-6

Flood protection along the Alpine Rhone river: Overall strategy and 3rd correction project

Tony Arborino
Roads and Watercourses Department, Canton of Wallis, Sion, Switzerland

Jean-Pierre Jordan
Federal Office for the Environment, Bern, Switzerland

ABSTRACT: More than 13,000 hectares of land in the Rhone Plain upstream from Lake Geneva are currently susceptible to flooding, including a high-density development zone covering an area of around 1,000 hectares. There are plans to invest more than 1 billion euros until 2030 in order to protect the Rhone Plain against hundred-year floods, and the agglomeration against thousand-year floods. The construction work will take several decades, and during this period a suitable response is to be made to the legitimate urgent calls for flood protection measures so that the development of the region can proceed without increasing the degree of risk. In order to secure the necessary level of acceptance for this major project, rapidly implementable structural measures, adapted land use planning and transparent communication about the degree of risk, will be essential.

Keywords: hazard maps; flooding; land use planning; anticipated measures; communication

1 INTRODUCTION

From its source to Lake Geneva, the Rhone runs through more than 160 km of highly-urbanized industrialized plain.

In recent years, the historic Rhone floods of 1987, 1993 and more particularly 2000, have clearly shown the hydraulic capacity and resistance limits of the Rhone dykes.

Over most of its course, the present layout of the Rhone is, in fact, unable to protect the plain against hundred-year floods. This is shown by an analysis of the current capacity of the river.

The lack of capacity is not due to sediment deposits on the river bed. The reality is quite the opposite. Since the middle of the last century, a general sinking of the Rhone bed, due to sustained gravel pit activity has been observed. The lack of hydraulic capacity of the River Rhone is compounded by the very poor condition of its dykes: between Brigue and Lake Geneva, half of them are dangerous, due to their poor stability with high risk of internal erosion and piping. The weakness is such that the dykes may break even before an overflow occurs. The present dykes date from the second Rhone correction (1930–1960). They were built over the initial dykes constructed during the first correction (1863–1884).

In Valais, more than 11,000 hectares of plain land are today under threat of flooding and more than 1,000 of these include the development zone that is already highly urbanized. Damage potential could amount to more than 8 billion euros. A third Rhone correction is therefore an absolute necessity to protect people and goods as well as to support the economic development of the plain.

The project approved in 2012 will ensure the entire plain's safety up to the hundred-year flood. The urban centers and major industries will be protected against far higher recurrence interval floods. However, even though there are plans to invest more than 1.5 billion euros

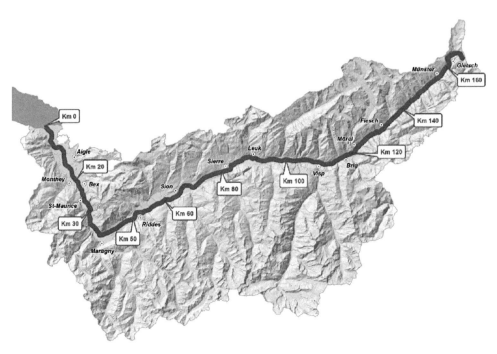

Figure 1.   The River Rhone and its watershed upstream of Lake Geneva.

by 2035 and priority measures are already underway, the plain will not be fully protected for several decades. Under such conditions the following questions arise:

1. What appropriate land-use regulations should be adopted during this transitional period before the project is completed, without considerably affecting land development?
2. What structural measures can be proposed to ensure the safety of the most sensitive areas as quickly as possible?
3. What are the problems related to risk communication and how should they be solved?

This overview therefore summarizes the experience gained in three areas: land-use planning, structural measures and communication, which take on a new dimension due to the scale of the project. The global approach developed will undoubtedly be valuable for the many future projects to renew the major flood-protection infrastructures that were inherited from the past and no longer meet current safety standards.

2   BRIEF FLOOD HISTORY

Prior to October 2000, the last major Rhone flood was in 1948, before the second correction. For many decades, no more major floods occurred until those of 1987 and 1993, which are the reason why the third Rhone correction was undertaken. Although the floods of 1987 and 1993 didn't cause much damages, they both reached the hydraulic capacity and resistance limit of the dykes. The situation was more critical during the October 2000 flood that caused a dyke to break and flooded large areas. The situation could have been even more critical if the zero degree isotherm had not suddenly fallen on the last day of the heavy rainfalls.

An analysis of the historic floods' hydrographs of 1987, 1993 and 2000 has shown that the meteorological conditions prevailing at the time of these rare events were relatively similar: several days of heavy rain arriving from the Southern Alps and driven by a warm and moist wind from the Mediterranean.

The Rhone floods are, however, also strongly influenced by rainfall distribution and, consequently, by the moment when the peak flows arrive from lateral rivers, causing a possible superposition of these flows. This makes the hydrological behavior of the watershed extremely complex and increases uncertainties.

The position of the 0°C isotherm above 3000 m and previous heavy rainfall, which are determining factors in the formation of these exceptional floods, are difficult to predict, especially if the effects of climate change are considered.

During these three floods, the peak flows significantly exceeded the historic values previously observed. Statistically, each of these floods was similar to the hundred-years flood obtained by statistical adjustment.

The second correction did not foresee flows of such magnitude. These historic events therefore resulted in revising the project flows upwards, integrating a 90% confidence interval to better reflect the uncertainties mentioned. The project flows were approximately 30% higher than the values adopted for the second correction (1930–1960).

3   MAP OF THE RHONE FLOOD HAZARD ZONES

Hazard zone maps are reference documents that take account of natural hazards and are used to develop land-use tools. They are also important for planning or developing regulations related to the protection of objects (building regulations) and other damage reduction measures.

The maps are accompanied by a report explaining the causes, course, magnitude, scope and probability of occurrence of natural hazards.

These maps cover practically the whole of Switzerland using identical hazard scale, independently from the natural hazard. High hazard is outlined in red, medium hazard in blue and low hazard in yellow. The map showing the zones at risk of flooding by the Rhone from its source to Lake Geneva was published in 2011. Previously, only an indicative map was available. The representation scale is 1:10,000 outside the construction zone and 1:2,000 within the construction zone.

Figure 2.   Example of a representation of the 2D model with the Rhone bed and the Rhone Plain (vertically distorted).

Figure 3. Hazard map of the city of Visp and its industrial area exposed to high or medium hazard covering most of the territory.

The hazard zones shown on the land-use maps are established on the basis of the hazard map calculated using a two-dimensional flood model taking into account the experience of recent floods and the specific characteristics of the Rhone plain. Its accuracy in calculating flood levels is within the range 10 centimetres.

The 2D model that is used links the Rhone and plain models to form a complete virtual model within which the hydraulic flows are calculated. It includes almost 1,300 cross-sections of the Rhone and 1,600,000 altitude points on the plain for a modelled area of 170 km^2, which constitutes a record density at this scale. It is the largest hydraulic model ever made in Switzerland.

This model also takes local characteristics of the plain into account thanks to many additional site visits and surveys. For example, highway, cantonal road or railway underpasses are taken into consideration, as floodwater may flow into them.

More than 11,000 ha of plain are therefore exposed to flooding. In particular, 1,055 ha of construction zone are located in the high-risk (red) zone and are potentially unsuitable for construction due to the usual natural hazard regulations. This very wide-ranging hazard situation, which concerns almost 100,000 people living on the plain and can potentially cause damages for more than 10 billion euros, is due to the situation of the Rhone with its rising water levels being on average 4 m higher than the plain and the risk of a dyke break or overtopping.

4   LAND-USE REGULATIONS

In view of the risks to people and goods, building permit regulations arising from Federal recommendations have been put in place and are consistently applied in Switzerland. They are valid for all natural hazards. In high hazard (red) zones, all building projects (new or conversion) are, in theory, prohibited.

It would be disproportionate and disastrous for the economic development of the plain if these land-use planning regulations were applied pending completion of the third Rhone correction planned for thirty years' time.

The canton then developed a complement to the hazard classification model recommended at Federal level. This model takes the Rhone flood development time into account and allows building in high hazard zones to take place under certain conditions. It has therefore been possible to lighten the Rhone high hazard regulations in certain areas.

In fact, a more detailed process analysis indicates two particular characteristics, in the case of the plain, which are likely to qualify how "hazard" is interpreted.

The Rhone plain flooding phenomenon is generally a slow process, except in the areas situated at the foot of the riverbank where it is much faster when the dyke breaks. In the event of flooding, the water level rises within an area bounded by the Rhone dyke and a downstream obstacle (often topographic). Due to the large surfaces of these areas, the water level rises slowly. Under such conditions, simple adjustments to the supporting structure of new buildings will allow them to withstand the static water pressures and avoid the sudden break that characterizes high hazard areas (sudden destruction of buildings).

In addition, floods occur only after several days of bad weather, several hours of rain and several hours of water runoff and routing water, passing through hydroelectric dam reservoirs whose retention effect is optimized by the MINERVE flood prediction and management tool. An analysis of these phenomena over time shows that if a critical flood level is set and defined by a ten-year recurrence time, there is a minimum period of six hours before the flood peak occurs. A comprehensive analysis of the areas, number of inhabitants and dwellings concerned has been undertaken by specialists in risk planning and organization in the event of a disaster. This analysis shows that if various prerequisites are met, such areas can be evacuated within this time interval. Communes will have to have prepared and regularly trained the entire evacuation operation (measures, times and resources).

In the light of these elements (no sudden destruction of buildings and the possibility of evacuation as flood waters rise) and if the conditions required for fast evacuation are guaranteed, areas at risk of high flooding (h > 2 m) not affected by high velocities (v•h < 2 m²/s) could be considered as regulated areas (as for average risk) and not as prohibited areas under territorial regulations.

This interpretation is, however, only possible in the exceptional case of slow hazard and if the following criteria are all met cumulatively:

1. The zone is already allocated for construction.
2. The construction zone is mainly built up.
3. The new buildings do not significantly increase risk.
4. The natural hazard is static flooding.
5. The new buildings are only authorized under condition that the regulations and/or usage restrictions minimize injuries to people or damage to important goods.
6. Emergency rescue operations ensure that the persons concerned can be evacuated in good time from dangerous areas and that the system is validated by the relevant cantonal authority.
7. The development zones are no longer within the red area when the third Rhone correction is completed (according to the development plan schedule).
8. No other natural hazard presents a strong threat to the area.

In addition, the land to be developed should not be subject to any other types of building restrictions (e.g. noise, groundwater protection, etc.).

5 PRIORITY STRUCTURAL MEASURES

A general project has been under development since 2005 and was put out for public consultation in 2008. This project combines widening the river with lowering the river bed and reinforcing the dykes to increase the hydraulic capacity without needing to raise the dykes. Raising of the dykes is in fact not a very robust concept (mainly due to the water being discharged into the plain water network and tributaries) and increases the residual risks when the hydraulic capacity is exceeded. The cost of the project is over two billion euros and it will take several decades to complete.

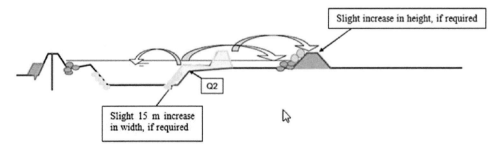

Figure 4.   First phase of the implementation of the measures.

That is why several priority areas have already been the subject of a detailed study carried out at the same time as the general project, in particular in Visp, an industrial and urban area, where work is currently in progress.

The operation to protect five priority zones in the midst of densely-populated built-up areas scheduled to be completed in 2025 is, however, not sufficient to meet the expectations of the authorities. Over and above the fears of citizens which the authorities must, in fact, address, land-use constraints have a significant effect on economic development and affect the entire plain. Solutions that can quickly be implemented must therefore also be proposed.

Rather than carrying out the work in phases, section by section, according to conventional practice, a study was conducted to propose partial measures that would have maximum efficiency, but be compatible with the preliminary project.

It entails the definition of several work phases in each section to achieve the protection objectives more quickly, whilst ensuring a reduction in investment costs, particularly in terms of materials management, and reducing the environmental impact of the works.

One of the solutions envisaged was to build, as a matter of priority, the inner dykes specified in the general project to manage residual risks and protect only inhabited areas. Permanent inner dykes are, however, not intended to deal with a Rhone dyke break scenario. If inner dykes had to be designed for this type of scenario, they would be higher than the Rhone dykes and have a significant influence and impact on the plain.

The general proposed phased implementation approach involves building the planned new dykes with the materials from the old dyke and strengthening the maintained dyke. The rock-fill or old groins will be removed to promote natural erosion and the blocks will be temporarily dumped at the foot of the new dyke.

By installing temporary gravel pits for 20–30 years and maintaining existing ones, the so created lack of bed load should facilitate lateral self-erosion in order to move towards the final transversal bed profile. This process must be regularly monitored and, if required, actively supported by mechanical material excavation if required.

The recommended measures must be tested on a first river stretch. If the results are conclusive, it should be possible to carry out the first stage described above to protect a large area of the plain against hundert years floods within 10 years, instead of the expected 30 to 40 years.

At the same time, the dykes will be equipped with safety weirs (fuse plugs) to manage overloads. It is, in fact, essential to prioritize comprehensive risk management throughout the flow range. This preference compared to measures that merely push back a flood threshold, without controlling the residual risk, or even exacerbate it, is one of the key elements of the study for the development of fast and efficient protection solutions.

6   RISK COMMUNICATION—ISSUES

The publication of the Rhone flood hazard zones highlighted the difficulty for the population to accept the idea of being in danger, in a society that aims at zero risk and looks for culprits to blame for every disaster.

The population and authorities will nevertheless first have to accept the fact that a risk exists and that it should be integrated into effective land management, otherwise building restrictions will not be understood and observed. Risk awareness and understanding must therefore be improved and the memory of risk recovered.

In fact, a comprehensive analysis of risk-related social dynamics (Fig. 5) shows that in the first land development phase on the Rhone plain, memory of risk was ever present and taken into account.

After major protection works had been carried out, the risk was forgotten and the land was developed without the risk being taken into account.

Publication of hazard maps and zones based on experience of major floods can now "materialize" the memory of risk which is taken into account in land development through land-use tools.

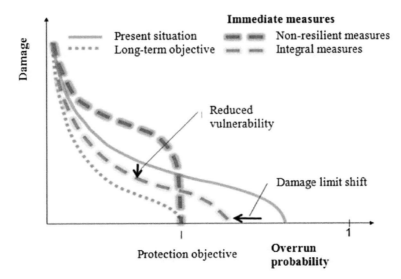

Figure 5. Effects of different actions on the risk.

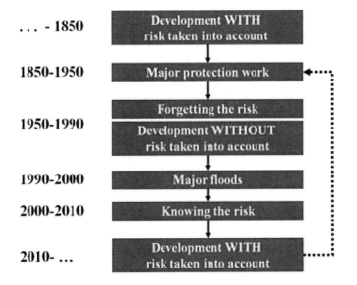

Figure 6. Evolution with time of taking into account the risk.

55

Finally, major flood protection projects must take the notion of residual risk into account once the work has been completed and map it, to avoid a loop and a repetition of the scenario after 1850.

After structural protection measures have been implemented, land regulations are vital to avoid a further increase in damage potential. Building regulations must therefore be maintained in areas of residual risk. According to current Confederation recommendations, the areas exposed to risk during an extreme event are mapped in streaked yellow, regardless of magnitude.

To better differentiate between land-use measures, the Canton of Wallis proposes to keep information regarding magnitude, using streaked yellow, streaked blue and streaked red to distinguish between the different zones. This distinction may be particularly useful when the first phased stage of the project is to be implemented.

If protection against rare floods is actually achieved after this first stage, it will only be possible to attain the objectives of robust measures when all the measures have been implemented, which should be reflected on the intermediate risk maps.

7 RISK COMMUNICATION—TOOLS

The main risk communication tool is the hazard zone map. It is available for public inspection and can therefore be viewed by the entire population. It is then transferred to zone allocation maps for information purposes (see Fig. 3).

Three degrees of risk (high—medium—low) are represented by three colors (red—blue—yellow) to simplify the way in which information is communicated by this type of map.

The risk information is presented in different ways to make it easier to view or understand.

The website and the GIS plug-in (www.vs.ch → Accueil > Transport, équipement et environnement > Protection contre les crues du Rhône > Danger et territoire) allow anyone who is interested to consult the on-line hazard zone map of the area concerned, print the required map and view essential information, such as the water level at a given point. The information can therefore easily be accessed by the whole population and is available for individual plots of land.

A fact sheet (A3, printed on both sides) giving an overall perspective of the hazard zones and answers to frequently asked questions has been published and distributed to the municipalities concerned.

Figure 7. Virtual representation of the hazards for improving the communication.

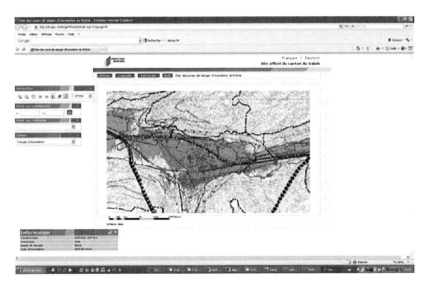

Figure 8.   Website of the Canton of Valais.

Figure 9.   Fact sheet on hazards due to flooding of the River Rhone addressed to municipal authorities.

A fact sheet in two languages distributed to the entire Valais population (150,000 copies) gives an overall perspective of the hazard zones and advice from the cantonal politicians concerned.

A brochure published on the occasion of the 10th anniversary of the flood of October 2000 also presented the processes involved and the importance of memory of risk.

Municipalities threatened by high risks are required to organize evacuation drills. If they had not specifically organized their emergency response drill and tested their structure and

57

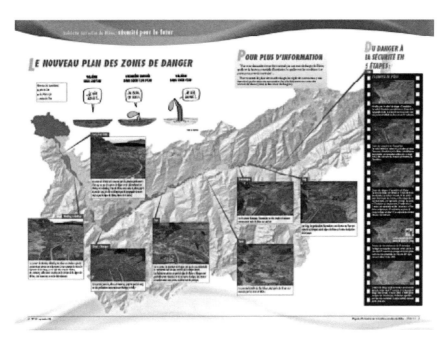

Figure 10.　Communication to households on hazards due to flooding of the River Rhone.

ability to evacuate the threatened population, the canton would respond negatively to requests to build in high hazard zones in this municipality.

At last, a hotline was organized and an ad-hoc working group set up to answer questions and, if necessary, identify new communication needs.

8　CONCLUSION

The project to develop the Rhone plain upstream from Lake Geneva is an example of the problems given by a critical hazard situation across a very large area that can only be resolved by work that will take several decades to complete. The fears of the inhabitants and the pressure of demands for protective measures to be quickly implemented must be addressed. In order not to compromise the acceptability of sustainable measures, other measures must be developed, whilst ensuring that full risk management is maintained.

Over the decades during which work on this large river development project will be carried out, land-use measures appropriate to the situation will be implemented to avoid any increase in damage potential, but without heavily compromising the development of the plain.

In addition to priority measures, the phased implementation of measures over virtually the entire course of the river should effectively and comprehensively reduce any risk in half the time envisaged to complete all the work.

This project is to date a unique case of large-scale adaptation of structures inherited from the large hydraulic developments of the nineteenth century and the beginning of the twentieth century. Taking account of new paradigms for this type of work in the field of flood protection therefore requires the development of special land-use tools, as well as participation and information tools. This experience will be very useful for future major embankment renewal projects in Switzerland or Europe that are similar to the international Alpine Rhine project upstream from Lake Constance, for which measures planning began a few years ago.

The work on priority measures in Visp that has been in progress since 2008 will also provide important insights into the work and allow the results of mathematical and physical models to be verified. Sustainable protection of an urban and industrial center with potential damage exceeding a billion euros is, however, already on the right track.

REFERENCES

Boillat,. J.L., (2005), L'influence des retenues valaisannes sur les crues, le projet MINERVE, Wasser Energie Luft, 11/12-2005, pp 317–324.
*Downloadable from the Valais canton web site: www.vs.ch → Accueil > Transport, équipement et environnement > Protection contre les crues du Rhône.
Département des transports, de l'équipement et de l'environnement du canton du Valais. Département de la sécurité et de l'environnement du canton de Vaud (2008). Rapport de synthèse du plan d'aménagement de la 3e correction du Rhône + (2010) Rapport intermédiaire sur l'information/la consultation publique de l'avant-projet du Plan d'aménagement (PA-R3) et du Plan sectoriel Vaud (PS-R3 VD) de la 3e correction du Rhône (mai à septembre 2008).*
Hernandez J.G., Horton P., Tobin C., Boillat J.-L., (2009) MINERVE 2010: prévision hydrométéorologique et gestion des crues sur le Rhône alpin, Wasser Energie Luft, 04-2009, pp 297–302.
OFEG (2001). — Directives protection contre les crues des cours d'eau. Office fédéral des eaux et de la géologie.
Petrascheck A., Hegg C., et al. OFEG, WSL (2000). Les crues 2000. Analyse des événements, cas exemplaires. Rapports de l'OFEG, série Eaux, no. 2, Office fédéral des eaux et de la géologie.
Roduit B., Arborino T., (2010). Crues 2000, Saillon se souvient.

Swiss Competences in River Engineering and Restoration – Schleiss, Speerli & Pfammatter (Eds)
© *2014 Taylor & Francis Group, London, ISBN 978-1-138-02676-6*

Flood control and revitalisation along the Aare river between Thun and Berne—experiences with recreational use and other conflicts of interest

F. Witschi
naturaqua PBK AG, Bern, Switzerland

B. Käufeler
IMPULS AG, Thun, Switzerland

ABSTRACT: The damage of the floods in 1999 and 2005 revealed the insufficiency of flood control on the Aare river between Thun and Berne. Riverbed-erosion has weakened the bank protection structures. It also causes a lowering of the ground water which leads to a loss of drinking water availability and induces changes in natural habitats. The vision of the *aarewasser* project: between Thun and Berne the Aare should flow through a river landscape in which natural dynamics are allowed in defined areas to support flood control, drinking water supply, ecology and recreational use.

The river and its banks are very attractive as recreation area. We have studied the behaviour of the visitors and how they can be motivated to act respectfully.

25 measures will change the river and the forms of land use in the affected areas. We have studied how and why interests collide and how these issues can be solved.

1 INTRODUCTION

1.1 *The Aare river between Thun and Berne*

Like the majority of the large Swiss midland rivers, the Aare river was canalised and embedded between closely spaced levees during the 19th century (Hügli 2007). Due to flood control it became possible and relatively secure to use the area behind the levees for forestry and agriculture. Various infrastructure facilities were built inside the levees or in the now protected area outside the levees. Railway lines, highways, drinking water wells and sewage treatment plants as well as their conduites are situated next to the Aare river. In former sidearms which are now cut off the development of wetlands of high ecological value could take place (Tiefbauamt Kanton Bern 2013).

1.2 *Need for action*

After the river training works and channelization was carried out, the Aare digs itself into the riverbed: on average the riverbed level sinks 1 centimeter per year. Consequences are a gradual disintegration of the bank protection structures as well as the lowering of the ground water level which in the long term leads to a loss of drinking water availability and a slow change in natural habitats. The loss of fauna and flora takes its course.

The floods of the years 1999 and 2005 revealed that the flood control of the Aare river between Thun and Berne is insufficient. Embedded in rigid levees the river claimed back its natural space and in some sections caused serious damage. Therefore since 2005 the flood control and revitalisation project *aarewasser* has been developed as a collaboration between the Canton of Berne, 18 communities and a "Schwellenkorporation". 25 flood and restoration

Figure 1.   From left to right: The Aare until the beginning of the 19th century; the Aare nowadays; the Aare after implementation of the *aarewasser* restoration project.

measures on the stretch of 25 kilometers are planned. The vision: Between Thun and Berne the Aare should flow again through a river landscape in which modern river engineering allows for more space and natural dynamics in defined areas. The project development is carried out in this spirit. The measures are expected to show positive effects on flood control, drinking water supply, ecology and recreational use.

1.3   *Urgent measure: Hunzigenau*

After the 2005 flood which caused serious damages including flooding of the highway and the Berne airport, the government of the Canton of Berne initiated the so called urgent measure Hunzigenau. Two side arms as well as a protection dam alongside the highway were built. Thus a first step was made towards the stabilisation of the riverbed, the flood protection of infrastructure and the revitalisation of river and wetlands. Since the completion of the revitalisation in 2006 the Hunzigenau is a popular destination for local recreation. It also serves as a example and is used to study the characteristics of the revitalised Aare river.

The work on the other planned measures is planned to start in winter 2016. After the public hearing of the overall project in 2009 more than 200 negotiations were lead. In spring 2014 the negotiations are now at the point where the project can be approved by the government.

2   CONFLICTS OF INTEREST

2.1   *Colliding interests*

The *aarewasser* project concerns various interests, some of which collide inevitably. This was expressed in a large number of objections (Tiefbauamt Kanton Bern 2013). For example:

Agriculture
Flood control and river engineering
Forestry
Ground water supply
Infrastructures
Nature conservation (amphibian spawning areas, alluvial zones, wetlands, fish spawning areas, fenlands, ...)
Protection of landscape and local character
Recreation
Settlements
a.s.o.

The interests are of different legal relevance. One of the main challenges of the project is to deal with the priorities of each interest according to the different local situations. Basically national law stipulates that flood protection measures must be combined with the restoration of the natural water course wherever possible. This requirement can be inconsistent with some of the interests (as mentioned above) and their related laws. If the intention is to respect every interest in the same way project blockades are inevitable. Two examples illustrate this point.

2.2 Example 1: Hydraulic engineering vs. drinking water supply

The drinking water wells in the community of Kiesen are the most important ones alongside the Aare between Thun and Berne (drinking water supply for the city and the region of Berne). They are located in immediate proximity to the Aare, as well as to the railway line and to the highway.

Basically construction measures within drinking water protection zones are not allowed. Nevertheless in the area of Uttigen/Kiesen (and therefore in the existing drinking water protection zone S2) various measures were planned.

The most significant riverbed erosion between Thun and Berne that has undermined flood protection structures occurs in the area of Uttigen. Hence the stabilisation of the riverbed has highest priority in this location. At the same time the high water levels in the region of Uttigen must be lowered in order to ensure long-term protection from floods due to ground water backlog. For logistic reasons a lowering of the high water level is only possible in combination with a reconstruction of the railway bridge (scheduled approx. for 2030). This reconstruction in turn is not possible without affecting the site of the drinking water wells.

As they would be carried out within the drinking water protection zone S2 or in the immediate area of the wells, all the planned measures to remediate the existing problems are not feasible legally: The project in this form is not environmentally compatible.

A solution was found by extending the considered perimeter. Within this greater area a definitive drinking water protection zone for a substitute drinking water well could be defined. By doing so the required conditions for the removal of some of the existing affected water wells were satisfied.

Triggered by this conflict between river engineering and drinking water supply the cantonal water authority initiated the elaboration of a *Masterplan Wasserversorgung Aaretal* (Ryser Ingenieure 2012). The masterplan coordinates the various projects concerning water supply and hydraulics respectively and points out dependencies and conflicts. Furthermore it identifies regions with higher need for coordination as well as synergies between projects.

2.3 Example 2: Fen protection vs. alluvial zone protection

The area called Chlihöchstettenau in the community of Rubigen is part of several nationally protected areas, e.g. the alluvial zone no. 69 „Belper Giessen", the fen no. 2635 „Chlihöchstettenau", the moor landscape no. 280 „Aare/Giessen", the amphibian spawning area BE574. Due to the levees built as bank protection structures fens developed in former alluvial zones. Out of this historical reason both habitat types with extraordinary nature

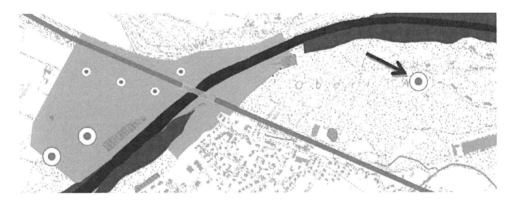

Figure 2.   Complex situation in Uttigen/Kiesen: dark line: Aare river; small dots: existing drinking water wells; large dots: replacement locations for the affected wells (arrow: well in extended perimeter); light grey surfaces: existing drinking water protection zone S2; dark grey surfaces: project measures; straight line: railway line with planned railway bridge.

values exist in the same area and are integrally protected. Now alluvial dynamics which are enabled by modern river restoration can threaten fens. We face a paradox situation.

Gradual riverbed erosion takes place also in the Chlihöchstettenau. Thus hydraulic measures must be taken here as well in order to provide for an equilibrium bedload level. What are the right measures in the light of the high and partly contradictory protection status of the area?

Legally the protection of fens (national constitution) stands above the protection of alluvial zones (national regulation). However by partially lowering the existing flood protection levee—as planned—one part of the Chlihöchstettenau will presumably be flooded on a regular basis and change into a more dynamic alluvial zone. The destruction of the fen surface can only be justified in the light of the overall project. The following arguments led to this decision: A) In the region alluvial zones are as rare as fens (and therefore as valuable). B) Within the project perimeter various zones will be revitalised which are of little ecological value nowadays. These revitalised zones are likely to offer habitats for species typically bound to alluvial zones and to fens. C) The transformation of the Chlihöchstettenau into a dynamised alluvial zone also seems reasonable in view of the fact that the Märchligenau (another nationally protected wetland area situated nearby) will be preserved and improved in quality. D) And finally another area in the valley of the Aare has lately been revitalised and was taken into consideration even though this project had nothing to do with *aarewasser*.

2.4 *Comprehensive project as success factor*

In order to overcome blockages within projects the authorities in charge are required to be open-minded and willing to compromise. They must be ready to leave defined paths with the objective of finding an agreement; this attitude requires courage and readiness to assume a risk. Of course such an open-minded position can only be taken within the legal frame.

The sum of existing conflicts is the same whether there are several small projects planned or whether they are summarised in one large-scale comprehensive project. However the solution of the conflicts is only possible within a large-scale perimeter.

What seems to be a reasonable consensus in the described examples was in fact the result of intense discussions and weighing of interests between the project team, the various concerned

Figure 3. Fen in the Chlihöchstettenau today. At the bottom of the picture the Aare, at the top the highway. (Picture: V. Maurer).

parties and authorities who were able to think in an interdisciplinary way. Without the large-scale consideration no solution would have been found for example 2 (Chlihöchstettenau). Only the overall view of all nature values alongside the Aare between Thun and Berne led the authorities to this holistic approach.

Thanks to the extension of the considered perimeter in example 1 an alternative to the existing, seemingly given situation about the drinking water wells could be found. The masterplan shows exemplarily how coordinated large-scale planning simplifies project development from the beginning. It is an important new perception for the traditional conflict between flood control engineering and drinking water protection.

3 RECREATIONAL USE

3.1 The Aare river—a magnet for recreation seekers

The Aare and its banks are very popular recreational areas. On nice summer days thousands of people spend their free time inside and alongside the Aare carrying out various activities. According to the visitor monitoring (Käufeler & Beutler 2012) the great majority of visitors seeks recovery and relaxation in nature. Other reasons for a visit are sporting activities, walking the dog and social company. As seen in the already remodelled Hunzigenau the Aare region is expected to be much more attractive as a recreational area after the flood protection and revitalising measures are carried out (Käufeler 2011).

3.2 What makes the Aare appealing to such an extent?

Which reasons lead to the attractiveness of the Aare and its banks? What makes them crowded with people on nice days? There are three main answers: First the high quality of the water and the adjacent habitats, second the easy access to the "water experience" and third the proximity to the densely populated agglomerations Berne and Thun.

3.2.1 Quality of natural habitats

The quality of habitats increases towards Berne. In the direction of Berne more and more ponds, open groundwater surfaces, alluvial forests, shallow swamps, reeds and low bank areas can be found. Width variation as well as flow variation increase, too. These qualities

Figure 4.   View of the revitalised Hunzigenau. (Picture: B. Käufeler).

are evidently appreciated by human visitors as the number of recreation seekers shows: it increases towards the city of Berne.

3.2.2 *Accessibility to the "water experience"*
Since the revitalisation of the Hunzigenau section and the improved access to the water the number of visitors has increased in a significant way which has some undesired side effects, e.g. the waste problem.

3.2.3 *Proximity of agglomerations*
The main part of the Aare visitors live in the agglomerations of the cities Thun and Berne as well as in the communities adjacent to the Aare. In total several hundred thousand persons live in an average driving distance of 10 to 15 minutes to the Aare.

3.3 *What visitor frequencies can be expected?*

High visitor frequencies along revitalised river sections raise fears in users and in public authorities. They are concerned with the following questions:

Do high visitor frequencies and/or specific kinds of visitor activities have an inacceptable influence on flora and fauna?
Does the amount of waste increase? Does it lead to unreasonably high maintenance expenses for communities?
Is there enough space for a peaceful coexistence of all the different user groups? Will there be conflicts? What kind of conflicts?

In order to gain reliable long-term answers to these questions and thus about the recreational quality the Canton of Berne has initiated the elaboration of a visitor monitoring (Käufeler & Beutler 2012). At defined times surveys about the following indicators are carried out:

Visitor frequencies (distinguished into the kind of activity)
Visitor behaviour (desired, tolerated, prohibited behaviour)
Knowledge and perception (e.g. knowledge of rules in the protection zone)

The surveys on the current status show that on nice summer days more than 5000 activities can be registered for certain river sections. Access to the area is gained by car (35%), by foot (28%), by bicycle (21%) and by public transports (16%) (Käufeler & Beutler 2012). Whereas in summer the dominating activities are boating and swimming, main spring activities are walking and biking, main winter activities walking and jogging.

Already today many recreation seekers use the Aare river and its banks. Yet the upcoming flood protection and revitalisation measures will attract even more. Of course the interest for the local recreational area is positive. But it can become a problem if the users don't adjust their behaviour accordingly.

3.4 *Conflicts alongside the river are diverse*

Conflicts in and alongside revitalised river sections can arise between persons and the existing partly protected nature values. They can also occur between different persons because of their different user interests. Thus conflicts are not only created by violations of the law but also by the spatial situation and by behaviour patterns that can be perceived as disturbing by the other visitors without being contradictory to law.

Before the revitalisation of the Hunzigenau the local maintenance employees in Rubigen collected 50 m³ of waste during the summer season. After the area was revitalised the amount of waste increased to 200 m³ for the same time span! After the community had taken some effort to reduce the direct access for cars the amount of waste could be reduced again. Due to provided mobile toilets, litter and excrement pollution of the area also decreased.

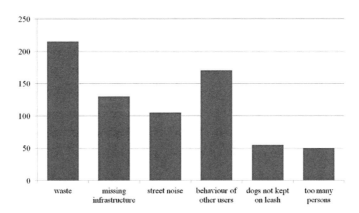

Figure 5.   Disturbances along the Aare as seen by the interviewed visitors. Basis: 500 interviews at different locations along the river between Thun and Berne.

From the sum of all recorded activities (N = 47,882) violations of current regulations and laws represent only 4% of the activities. These are mainly disregard of traffic ban or disregard of the duty to take dogs on a leash.

3.5 *Solution strategies*

How can we achieve a recreational use of revitalised river sectors with as little conflicts as possible? The Canton of Berne launched the elaboration of the concept of visitor guide and information (Gerber et al. 2012) which proposes solutions to this question. Because of the very diverse river area different solution strategies must be applied (Käufeler & Gerber 2008).
   Basically in the *aarewasser* project four strategies are implemented:

1. Identifying and considering the (diverse) ecological sensitivity of the individual sectors
2. Bundling the offered infrastructure with the aim to guide the visitors by these infrastructures
3. Establishing a homogeneous information and guidance system
4. Supervising and advisory

First the identification of the different degrees of quality and sensitivity of all the sectors takes place. The further implementation of the solution strategies will happen by stages. A set of measures will be defined adapted to the respective sector. The visitors' guidance will be carried out by the means of infrastructure offers and prohibition. We consider awareness raising and a legal foundation important factors to achieve the requested behavior. Three planning instruments will be applied in order to implement the regulations:

Revision of the conservation area resolution and the associated plan, addition of new protection regulations (in particular restricted walkways and a dog leash obligation in protected areas)
Communal planning of the riverbank protection (regulations about fire places, walkways, information sites)
Forest road plan (driving rights and ban of driving within forests)

3.6 *Challenge*

Revitalised river sectors attract more recreation seekers. This can lead to conflicts between people and nature or between different people looking for leisure activities. The Canton of Berne, responsible for the *aarewasser* project, has identified the need for action in the topic field of recreational use and has created suitable instruments and solution strategies. In a large project like *aarewasser* a global approach is possible and—moreover—essential!

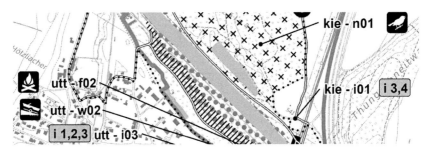

Figure 6. Detail of the concept plan *visitors' information and guidance* (Gerber et al. 2012). Grey surface: Aare and ponds. Dotted: Future river basin area. Cross-hatching: Areas where nature has priority. Lined: Areas with expected high visitor frequency.

The knowledge of the visitors' perception and behavior is a factor of success. Because of the great implications of leisure activities along the river a flood control and revitalisation project is hardly sustainable without considering the recreation aspect. The interest in the aspect of recreational use must not be restricted to the planning phase. Only the long-term observation allows for required adjustments with respect to recreation and nature quality. The greatest future challenge will probably be the implementation of reasonable, effective and at the same time affordable measures.

4 CONCLUSIONS

The amount of conflicts of interest in river restoration projects in the densely populated Swiss river valleys can only be approached on an overall, large-scale view. A comprehensive project (like *aarewasser*) offers the chance to find solutions that aren't possible in small-scale projects. Well-structured balance in bedload equilibrium, between different nature values, between nature and recreational use, a.s.o. can hardly be achieved in small-scale areas, but only on a regional basis. Only in the context of a comprehensive project with enough broadness in space and contents, blockages can be solved by means of broad-minded solutions and weighing of all interests—within the legal frame!

All the project stakeholders must have the readiness to take on an overall view and to consider the numerous (partly contradictory) project aspects from all view angles to allow a transdisciplinary approach. The same is valid for communities. They are forced to think in a larger scale than their community borders. Sometimes community representatives are over-taxed by the large-scale project approach. This makes it all the more important for the project management to guarantee that all the concerns of the various players are taken seriously and will be integrated in the overall consideration. A sufficient span of time is important.

The propagated large-scale, transdisciplinary and pragmatic thinking is the challenge of the modern river engineering and restoration.

REFERENCES

Gerber, V., Käufeler, B., Reinhard, K. 2012. *Besucherinformation und -führung -Konzeptplan und Konzeptbericht*. OIK II Kanton Bern.
Hügli, A. 2007. *Aarewasser. 500 Jahre Hochwasserschutz zwischen Thun und Bern*. Bern: Ott Verlag.
Käufeler, B. 2011. *Wenn die Aare ruft*. Quartalsthema *aarewasser*. OIK II Kanton Bern.
Käufeler, B. & Beutler, R. 2012. *Besuchermonitoring aarewasser*. Kurzbericht. OIK II Kanton Bern.
Käufeler, B. & Gerber, V. 2008. *Nutzen und Schützen in Einklang bringen*. Quartalsthema *aarewasser*. OIK II Kanton Bern.
Ryser Ingenieure 2012. *Masterplan Wasserversorgung Aaretal*. Erläuterungen zum Übersichts- und Terminplan. Technischer Kurzbericht (bereinigte Version 1).
Tiefbauamt Kanton Bern 2013. *Technischer Bericht mit Kostenschätzung*.

Swiss Competences in River Engineering and Restoration – Schleiss, Speerli & Pfammatter (Eds)
© 2014 Taylor & Francis Group, London, ISBN 978-1-138-02676-6

Flood characteristics and flood protection concepts in the Reuss catchment basin

Peter Billeter, Matthias Mende & Jolanda Jenzer
IUB Engineering Ltd., Berne, Switzerland

ABSTRACT: The paper summarizes both a conceptual study on flood behaviour in the Reuss catchment area and various flood control projects along the Reuss and its main tributary, the Kleine Emme. To analyse flood behaviour in the Reuss system extended flood routing computations were carried out using a hydrological model. Possible flood control and flood protection measures along the Kleine Emme and the Reuss downstream of Lake of Lucerne are explained and future perspectives of flood management in the Reuss catchment basin are pointed out.

1 INTRODUCTION

1.1 *The Reuss catchment basin*

The catchment area of the Reuss reaches from the central Alps of Switzerland to the so-called "Wasserschloss" where the Reuss confluences with the two rivers Aare and Limmat. The Reuss system drains an area of about 3'425 km² and consists principally of three sections:

1. The catchement area of the Lake of Lucerne, which has a lake surface area of 114 km² and efficiently damps out floods from the upstream tributaries,
2. The catchment area of the Kleine Emme, the latter dominating flood characteristics downstream of the Lake of Lucerne and
3. The area downstream of the confluence of Kleine Emme and Reuss to the Wasserschloss.

Figure 1 shows the situation and gives an overview of the catchment basin. Furthermore, the location of the gauging stations for discharge and precipitation is given.

A closer focus has to be put on the confluence of the Reuss and the Kleine Emme right downstream of Lake of Lucerne (Figs. 1, 3 and 4): The outflow of the lake of Lucerne is controlled by the so-called "Reusswehr" which is basically a needle weir. The weir was refurbished during the last years. An additional weir field with a flap gate was added laterally in order to increase the discharge capacity of the lake outflow.

Furthermore, the capacity of the Reusswehr is strongly influenced by the discharge of the Kleine Emme at its confluence with the Reuss. This confluence is called "Reusszopf" and lies about 2.7 km downstream of the Reusswehr. In case the discharge of the Kleine Emme is increased, the water level at the Reusszopf increases such that the Reusswehr is partially submerged and the outflow of the lake gets reduced. At a lake level of 434.00 m a.s.l. and a flood discharge of 700 m³/s in the Kleine Emme for instance, the outflow of the lake of Lucerne is reduced by nearly one third (i.e. about 90 m³/s). Furthermore, the regulation possibilities of the Reusswehr are limited since the needles can be set and removed only up to lake levels lower than 433.7 m a.s.l and 434.0 m a.s.l., respectively.

1.2 *Flood generation and flood characteristics*

The analysis of recent flood events in the Reuss basin showed that the occurrence of flood downstream of the "Reusszopf" is mainly a consequence of flood in the Kleine Emme, the

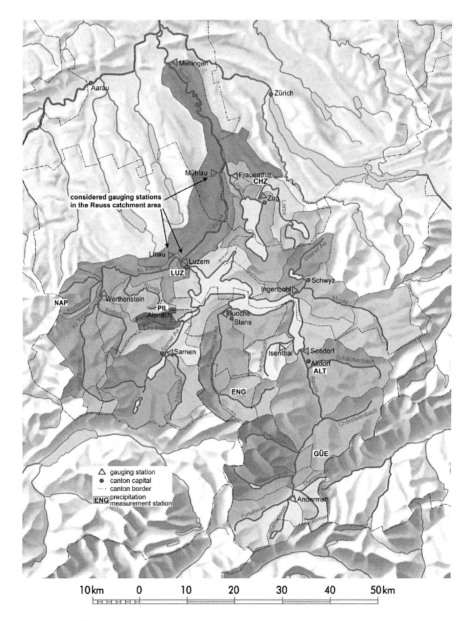

Figure 1. Map of the Reuss catchment area with the gauging stations for surface water level and discharge as well as for precipitation.

Kleine Emme being a river with very quick watershed response and therefore rapid swelling and accentuated peak discharges. However, the contribution of the Kleine Emme is a necessary but not a sufficient condition for floods further downstream. To produce flood, a significant contribution of the Reuss outflowing of Lake of Lucerne is necessary. Taking this into account two types of characteristic flood events can be identified for the Reuss downstream of Lucerne:

a. Flood events in spring and early summer which are dominated by snow melting and consequently high water level of the Lake of Lucerne. For this situation already moderate flood events in the Kleine Emme might be sufficient to produce flood downstream of the Reusszopf. Peak discharge and peak lake water level usually do not coincide (see Table 1: 1999 and 2004).

Table 1. Analysis of flood events, peak discharge and temporal offset of peak discharge of the Kleine Emme upstream of the confluence with the outflow of the Lake of Lucerne (Reuss LU), Reuss AG indicates the combined discharge of Reuss and Kleine Emme further downstream in the canton of Aargau.

Flood event	Q_{max} [m³/s] Kleine Emme	Q_{max} [m³/s] Reuss LU	$\Delta t\,(Q_{max})$ [h]	$\dfrac{Q_{max,Emme}}{Q_{max,ReussAG}}$ [%]	Q_{max} [m³/s] Reuss AG
Aug 1978	472	274	48	**74**	639
Mai 1999	332	410	240	**47**	711
Jun 2004	340	327	273	**57**	597
Aug 2005	700	473	55	**77**	839
Aug 2007	559	316	40	**76**	740

b. Extreme flood in the Kleine Emme catchment area which occur mostly in summertime and after an extended period of bad weather with a sufficient amount of precipitation to produce saturized soils. Since the catchment areas of both the Kleine Emme and the Lake of Lucerne lie side by side and parallel to the west—east alignment of the Alps, significant flood discharge will occur in both basins during temporally and spacially extended rainfall periods. For this situation however, the retention of the Lake of Lucerne hinders coincidence of the flood peaks from the Kleine Emme and the outflow from Lake of Lucerne. As can be seen in Table 1 the peak discharge of the Kleine Emme and the peak level of the Lake of Lucerne have a time gap of about 2 days which is longer than the duration of the flood peak of the Kleine Emme.

Most feared is a combination of the two flood types, i.e. an early summer flood with long enduring and spacially extended rainfall occurring simultaneously with extreme water level of the Lake of Lucerne caused by snow melting.

2 METHOLOGY

2.1 Flood routing modelling and model calibration

To analyze flood behaviour in the Reuss system extended flood routing computations were carried out using a hydrological model developed by the Laboratory of Hydraulic Constructions (LCH) at the Swiss Federal Institute of Technology in Lausanne (EPFL) (Hernandez et al., 2007). The model was calibrated using the flood events of May 1999 (spring flood combined with snow melt) and August 2005 (summer flood caused by extended rainfall).

On the one hand, the computation results helped to understand flood generation mechanisms in the catchment area. On the other hand the model was used to determine the effect and benefit of flood protection measures regarding the decrease of flood peak discharge downstream of the confluence of Reuss and Kleine Emme (i.e. downstream of the Reusszopf). As an example Figure 2a shows the comparison of measured and computed time series of flood discharge for 2005 summer flood. Figure 2b shows the time series for the same flood type but with excess peak discharge ($1.3 \times Q_{dim}$ with Q_{dim} = design discharge, see Sect. 3) comparing the discharge without and with flood control measures as described in Section 3.

2.2 Efficiency of flood protection measures and assessment of cost-benefit ratio

As mentioned above, the hydraulic effect of flood control measures was determined by a hydrological model. The data of discharge and water level (i.e. the reduction of peak discharge and peak lake and flood plain level, respectively) where coupled with flood damage costs collected during earlier flood events. Flood damage costs were plotted as a function of either water levels or discharge. The combination of flood damage cost and the effect of flood control measures upon discharge and water level allowed for both a quantitative flood

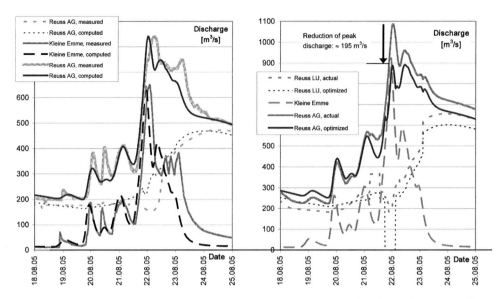

Figure 2. (a) Model calibration, comparison of measured and computed discharge for the 2005 flood (HQ_{2005}). (b) Hydraulic efficiency of flood control measures: Optimized reservoir management for excess flood of $1.3 \times Q_{dim} = 1.3 \times HQ_{2005}$.

risk assessment of the Reuss catchment area and the evaluation of the cost-benefit ratio of flood control measures.

3 FLOOD PROTECTION MEASURES

A great variety of both flood control concepts and specific flood protection measures were either developed or gathered from previous studies and investigations. Figure 3 gives a brief overview. Some of these measures will be described in the subsequent sections.

3.1 *Improved water level management of Lake of Lucerne*

3.1.1 *Improvement of lake regulation and capacity of outflow control structure*
The outflow control structure of Lake of Lucerne, the so-called "Reusswehr" (Fig. 4), was refurbished and extended in the last years and new rules for the lake management were installed. Nevertheless, flood control measures involving an improved water level management of Lake of Lucerne are suggested. The implementation of these measures will take some years or even decades since it requires a modification of the recently commissioned lake management rules.

Furthermore, the construction of an auxiliary weir upstream of the existing Reusswehr was investigated and preliminarily designed (Figs. 4 and 5a). The concept consists of a rubber weir hidden right downstream of the brigde "Seebrücke" in order not to disturb the historical townscape of Lucerne.

3.1.2 *Bypass tunnel for the Kleine Emme*
Since flood peaks in the Kleine Emme are short (≈12 h) it was sought so deviate the Kleine Emme into the Lake of Lucerne by means of a bypass tunnel. This idea was conceived already in the 19th century. Figure 5b shows possible layouts of the bypass tunnel. Var. 1 corresponds to the 19th century solution and was found to be not feasible taking groundwater interaction and the geology into account. Var. 2 was studied in more detail. A free flow tunnel with a diameter of 6.5 m and a deviation capacity of about 200 m³/s was designed including an

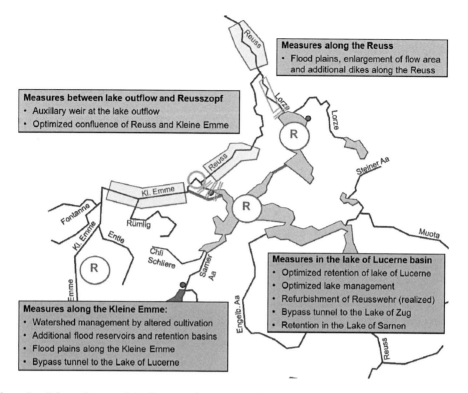

Figure 3. Schematic map of the Reuss catchment area with the investigated flood protection measures.

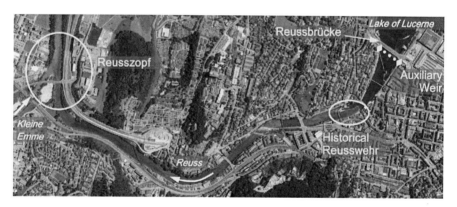

Figure 4. Aerial view of the outflow of Lake of Lucerne and the Reusszopf (confluence of Reuss and Kleine Emme) with the position of the historical weir and the suggested auxiliary weir.

intake side weir along the Kleine Emme and the outflow structure into the Lake of Lucerne and both the feasibility and the cost of this bypass solution were assessed.

3.2 *Basic concepts of flood protection and river training along the Kleine Emme and Reuss*

3.2.1 *Flood protection*

The design flood Q_{dim} at the lower part of the Kleine Emme is 700 m³/s, Q_{dim} of the Reuss downstream of the confluence with the Kleine Emme amounts to 840 m³/s. For both design floods the annual probability of occurrence is about 0.01 (100-years flood). The design of

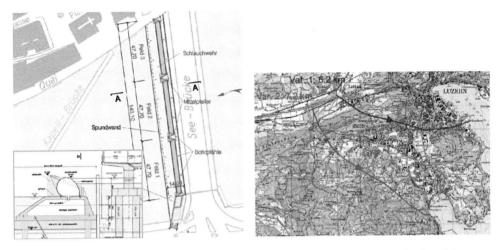

Figure 5. (a) Plane view and cross section of suggested auxiliary weir at outflow of Lake of Lucerne. (b) Suggested layout of a bypass tunnel between the Kleine Emme and the Lake of Lucerne.

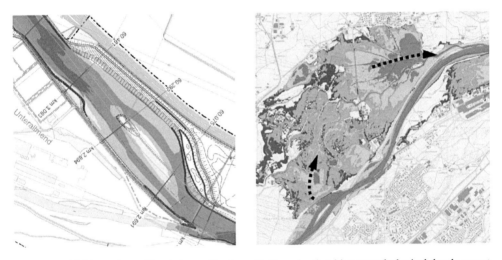

Figure 6. (a) Plane view of local river widening with flow depth taking morphological development into account. (b) 2d shallow water equation simulation of a flood plain artificially flooded during excess flood.

the river bed (i.e. width and bank height) accounts for a safety margin added to water level at design flood allowing a flood up to $1.3 \div 1.5 \times Q_{dim}$ to discharge without overtopping.

Basic concept and first priority of the protection measures is the enlargement of the flow area of the rivers Reuss and Kleine Emme. If necessary the enlargement is combined with new or reinforced dikes. To enhance the detention effect of enlarged river beds, flood plains and flood corridors are defined and limited by secondary dikes. These plains and corridors will be flooded either artificially or naturally during excess flood discharge. Figure 6a shows a typical plane view of a local widening with characteristic flow pattern for average summer discharge.

Figure 6b is an example of a 2d shallow water equation simulation of a flood corridor which is flooded artificially to handle excess floods bigger than $1.3 \times Q_{dim}$. Buildings with higher safety requirements inside the flood corridor will be equipped with individual tailor-made protection gear.

3.2.2 River training: Enhancing morphological variability of river beds

For environmental purpose river engineering measures have to be carried out in accordance to the Swiss Federal Law should improve morphological variability of the river bed. Along the Kleine Emme and the Reuss two concepts are used:

I. Outside of building zones local river widening will be built at suitable sections of the river in order to both stabilize the river bed against vertical erosion and enhance structural variability.

II. Within densely populated urban areas where no space is available for the widening of the river bed, instream river training measures will be applied within the existing river bed (i.e. micro groynes in the river bed and structural elements as wood and blocks along the banks; see e.g. Mende, 2012). Figure 7 shows the situation and some typical details as planned for the Kleine Emme.

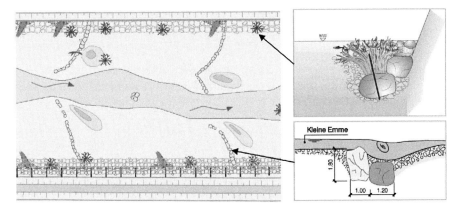

Figure 7. Increasing the morphological variability of the Kleine Emme in urban areas by instream river training measures.

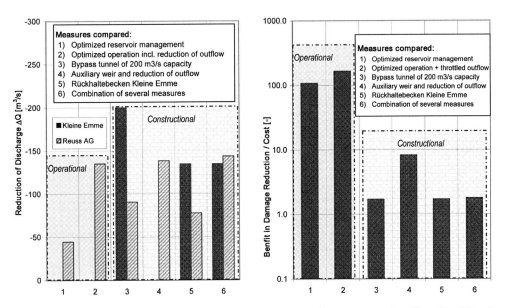

Figure 8. (a) Hydraulic efficiency of flood control measures. (b) Inverse cost-benefit ratio of flood control measures (Benefit in damage reduction versus cost of measure).

75

4 EFFICIENCY AND COST-BENEFIT RATIO OF FLOOD PROTECTION MEASURES

As mentioned in Section 2, a combination of flood routing computation and flood damage function was used to assess both the hydraulic efficiency and cost benefit ratio of flood control measures described in Section 3.

Figure 8a compares the hydraulic efficiency i.e. the reduction of peak discharge of various measures. Two groups of measures can be distinguished: (a) operational measures involving optimization of reservoir management and (b) constructional measures as e.g. an auxiliary weir, additional retention basins or a bypass tunnel.

Figure 8b shows the cost-benefit ratio of these measures. Generally, the hydraulic efficiency of constructional measure is better than the efficiency of operational measures. However, if the cost-benefit ratio is considered, operational measures are always better judged and consequently much more efficient with regard to the benefit of capital invested into flood protection measures.

5 CONCLUSION

Flood characteristics of the Reuss catchment area were analyzed and flood protection measures were developed in order to reduce the flood risk downstream of the Lake of Lucerne. The hydraulic efficiency as well as the cost-benefit ratio of the measures were assessed by combining flood routing computation with data of damage costs as a function of flood discharge and water level, respectively. The study demonstrated that operational measures such as optimizing the water level management of Lake of Lucerne tend to have a better cost-benefit ratio than constructional measures.

Since the implementation of the measures developed in the present study will require an extended process to gain authorization and political acceptance, river training and flood protection measures along the rivers Kleine Emme und Reuss are pursued in first priority. These measures involve extended enlargement of the flow area and flood corridors outside of the river bed, both aiming at save and slightly retarded discharge up to and even above the respective design flood discharge. Some of the major flood protection projects along the Kleine Emme and the Reuss have already reached the final design stage. The projects along the lower part of the Kleine Emme are authorized and construction started at the Reusszopf in 2013.

ACKNOWLEDGMENT

The conceptual study and the planning of the flood protection measured presented in this papers were carried out by commission and under the guidance of the Traffic and Infrastructure Authority (vif) of the Canton of Lucerne. Support and permission to publish are greatly acknowledged.

REFERENCES

Hernandez, J.G.; Jordan, F.; Dubois, J.; Boillat, J.-L. 2007. Routing System II—Flow modelling in hydraulic systems. Laboratory of Hydraulic Constructions, Ecole Polytechnique Federale de Lausanne, *LCH Communication 32*, Lausanne.

Mende, M. 2012. Instream River Training—Naturnaher Flussbau mit minimalem Materialeinsatz. *Korrespondenz Wasserwirtschaft*, 5(10), 537–543.

Swiss Competences in River Engineering and Restoration – Schleiss, Speerli & Pfammatter (Eds)
© 2014 Taylor & Francis Group, London, ISBN 978-1-138-02676-6

Innovative measures for management of bed load sediment transport: Case studies from alpine rivers in western Switzerland

G. de Montmollin
STUCKY Ltd., Renens, Switzerland

A. Neumann
STUCKY Ltd., Renens, Switzerland (Currently: SDC, Berne, Switzerland)

ABSTRACT: During the last decades, the answer to the problems caused locally by sediments along small alpine rivers, especially on their alluvial fans, has been the construction of sediment dumps, combined with regular sills in the river to prevent downstream erosion. This paper presents innovative measures that allow maintaining the sedimentary dynamics of the river for average floods, so they ensure an optimal protection against natural hazards and do not reduce the quality of the river in terms of fish migration, fish reproduction, or the aesthetic quality for local residents.

1 INTRODUCTION

Problems caused by occasional large sediment deposits in riverbeds have historically been resolved by the construction of sediment dumps. Such a solution is problematic from an environmental perspective and often causes erosion problems downstream. Alternatives are to work in the catchment area by limiting the sediment intake and the transport capacity of the river, for example by the construction of weirs or bank riprap. Such solutions are usually expensive in terms of investment and maintenance.

The purpose of this paper is to present innovative measures that have been realized on alpine rivers in western Switzerland between 2007 and 2012, and one on-going project. The first project is the sediment retention area located upstream of the city of Bex on the Avançon River. This basin has a defining characteristic of being located parallel to the river, hence the need to spill the sediment over a lateral weir. The second project is the sediment retention section located upstream of the city of Châtel-Saint-Denis on the river Veveyse de Châtel. The concept is limited to a quite small intervention which allows an important increase of the sediment retention capacity of an active sedimentation section. The third project (ongoing) is the correction of the Baye de Clarens in Montreux to increase the sediment transport capacity up to Lake Geneva and the modification of the mouth to allow deposition of the inflowing sediment volume without generating an increase in the risk of flooding.

The three projects are all upstream of or inside densely urbanized areas and on rivers with high ecological value. In this kind of setting, both the robustness of the proposed solution in relation to the natural hazards and the quality of the landscape integration of the solution are very important. For the proposed solutions, (1) the required maintenance of the structure is very limited and large-scale maintenance is necessary only in case of important flood events; consequently, disturbance of local residents is very low, and (2) the projects do not reduce the sedimentary dynamics of the river for average floods, so they do not reduce the quality of the river in terms of fish migration, fish reproduction, or local residents recreational activities. On the contrary, recreational and ecological values of the sites are being improved.

2 LATERAL SEDIMENT RETENTION AREA ON AVANÇON RIVER

2.1 Introduction

The sediment transport capacity of the Avançon River through the town of Bex is insufficient during floods. In October 2000, during a 20 years event, the water level rose to danger limit due to the high amount of alluvium accumulated in the river channel. After the flood, a map of natural hazards and a flood protection plan was drawn up. The plan encompasses the increase of the hydraulic capacity of the river through the town and the construction of a sediment retention area in parallel to the river, upstream of the town.

The sediment retention area includes an innovative concept of the derivation system, composed of a sill with orifices in the stream and a lateral weir. For this reason, it was required to test and validate its functioning on a physical model. This validation has been realized by the Laboratory of Hydraulic Constructions (LCH) of the EPFL on physical model at a scale of 1:30 (concerning the physical model refer to Ghilardi et al. 2012). Following the validation of the concept, it was realized between September 2010 and May 2011.

2.2 Catchment area and location

The main characteristics of the Avançon river catchment upstream from the sediment retention area are given in Table 1. The sediment retention area is located at the apex of the alluvial fan of the Avançon River. Upstream of the alluvial fan, the slope of the river is above 8%. On the alluvial fan the slope decreases slowly down to 1.5%. Figure 1 shows the longitudinal profile of the Avançon River and its two main tributaries.

The catchment area is mountainous. It includes very long segments with unstable banks and tributaries with regular mudflows. The volume of sediment which could be eroded by the river is notable, almost unlimited. The volume of sediment which arrives on the alluvial fan is given by the sediment transport capacity of the river (based on Jäggi 1992 and

Table 1. Main characteristics of the catchment upstream from the sediment retention area.

Surface (km²)	Perimeter (km)	Elevation			Average discharge (m³/s)	Flood discharge		
		Max. (masl)	Mean (masl)	Min. (masl)		30-year (m³/s)	100-year (m³/s)	300-year (m³/s)
77.3	42.0	3210	1747	469	3.9	60	79	96

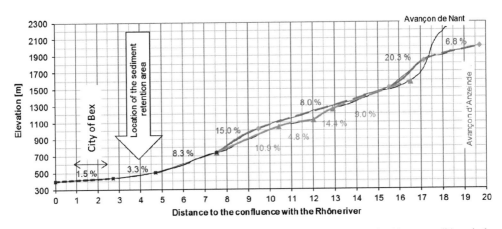

Figure 1. Longitudinal profile of the Avançon River and its two main tributaries (Avançon d'Anzeinde and Avançon de Nant).

Jäggi 1995). Consequently for each major flood, important deposits are observed through the urban area.

2.3 *Project presentation*

To avoid destabilizing the river bed and improve its aesthetics, the sediment retention area was not designed as a "classic" sediment dump in the flow path. In fact, numerous observations on such dumps show that they retain most of the sediment carried downwards, regardless of the discharge and not only in times of important floods. Downstream from such dumps, significant erosion of the bed is often observed, leading to a destabilization of the banks and sometimes to the ruin of the existing sills. This erosive dynamic is the direct consequence of the sediment deficit resulting from the dump upstream. The function of the sediment retention area should therefore be not to stop all the sediment transiting in the river, but only the sediment which exceeds the downstream transport capacity of the river. For the Avançon River, the sediment retention area is designed to begin to function for a flood with a return period ≤ 20 years. For such events, the river can provide a quantity of sediment of around 17,000 m³ to 25,000 m³. Relatively clear criteria have been defined for the design of the spreading area, as reported in Table 2.

The sediment retention area is designed for a 100-year flood and offers a storage capacity of about 12,000 m³. In addition, a volume of about 8,000 m³ can be stored in the bed of the Avançon on the segment located along the retention area. In a 100-year event, the residual sediment downstream of the retention area should correspond to the capacity of the Avançon through the city of Bex.

From a functional point of view, the sediment retention area is characterized by three components (Fig. 3): a derivation unit, a sedimentary basin and a downstream regulation

Table 2. Design criteria for the sediment derivation and retention system of the Avançon.

Frequency of the event (return period)	Peak flow (m³/s)	Bed load volume carried by the Avançon river (m³)	Bed load volume stopped in the sediment retention area (m³)	Bed load volume stopped in the Avançon river due to the retention area (m³)
Yearly	26	Undefined	0	Undefined
20 yrs	45	17,000	6,000	~6,000
100 yrs	79	25,000	12,000	8,000
Extreme	120	~50,000	Same as above	Same as above

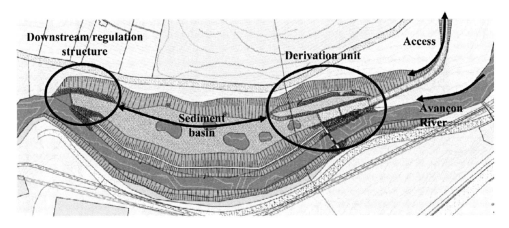

Figure 2.   Topographic plan view of the project.

Figure 3.   Photograph of the derivation unit, from upstream.

structure. The derivation unit has the aim to control the portion of the flows, solid and liquid, which are discharged into the basin. The retention pond or spreading area is the surface where the bed material will settle. Finally, at the downstream extremity of the basin, the geometry of the structure is expected to optimize the storage capacity of the sedimentary basin and the return of the liquid flow back to the Avançon.

The derivation unit (Fig. 3) consists of a sill with orifices in the Avançon River and a lateral weir on the right upstream side. The sill, which is around 9 m wide, includes two orifices with a width of 3.20 m each capped by a wooden beam. The geometry induces a contraction of the flow and causes a rise of the water level during the flood. The form of the lateral weir plays in this case a fundamental role for the amount of derived sediments.

Grooves are placed on the side of each orifice. Therefore, the capping beams are adjustable in height, allowing a possible subsequent modification of the geometry of the derivation unit.

2.4   *Operation during floods*

During a common flood (up to the yearly flood), the river flow stays in the river bed and bypasses the basin; the basin is slightly flooded at the downstream end. When the discharge increases above that of a common flood, water, and later on water with sediment, will be released into the basin and the basin will progressively become filled, as illustrated in Figure 4.

a. During a flood with a relatively low return period or at the beginning of a major flood:
   – The sill generates an increase of the upstream water level, which slows down the flow.
   – If the flood carries bed load, sedimentary deposits are created upstream of the sill.
   – The Avançon River overflows into basin.
b. During a major flood (Figure 4-b):
   – The sedimentary basin is flooded, first with water slightly loaded with sediment, then with increasingly loaded water.
   – The sediments settle in the upper part of the basin.
   – In the Avançon River, the water discharge reduction outpaces the bed load reduction. Consequently, deposits are also generated in the Avançon downstream of the sill.
c. Towards the end of a major flood (Figure 4-c):
   – The basin is full of sediment (3 to 5 m thick).
   – On the basin side, the derivation unit is also full of sediment.
   – Due to the widening of the channel upstream of the basin, the main part of the discharge flows in the Avançon.
   – The sediment in the Avançon is partially eroded.

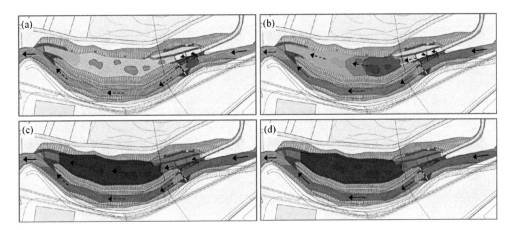

Figure 4.   Operation phases of the sediment retention area of the Avançon River during a major flood. (a) at the beginning of overflow towards the basin; (b) during the flood; (c) towards the end of the flood; (d) after the flood.

d. After a major flood (Figure 4-d):
  – The basin has to be emptied and ponds recreated.
  – In the river, the sediment must be removed upstream of the sill.
  – Downstream of the sill, the sediment will be naturally removed (this process can be relatively long).

2.5   Intermediate conclusion

The sediment retention area realized on the Avançon upstream of the city of Bex has no impact on the bed load dynamics of common floods and of major floods with low bed load. It therefore has the following advantages:

– Conservation of a partial bed load dynamic during a major flood and therefore limitation of the erosion processes downstream.
– Minor impact on the environmental quality of the river (conservation of the longitudinal continuity and bed load dynamics).
– Maintenance only in case of major floods.

3   SEDIMENT RETENTION AND BREATHING SECTION, VEVEYSE DE CHÂTEL

3.1   Introduction

The hydraulic capacity of the Veveyse de Châtel mountain river through the city of Châtel-St-Denis is insufficient during floods. Furthermore, the capacity can be reduced and the risk of overflow heavily increased with bed load deposits and bridge obstruction by floating material.

   Based on bed load calculations (using the guidelines from GHO, 1999) and the natural hazard map, it has been decided to secure the river through the city. For this purpose, a concept of protection against floods has been defined. This concept includes the increase of the hydraulic capacity of the river through the city, the creation of small dikes outside the river to limit the spread of flows and the construction of a sediment retention section, upstream from the city. The sediment retention section is composed of sills which limit the hydraulic capacity of the river bed and generate an overflow. This scheme is described hereafter.

3.2 Catchment area and location

The main characteristics of the Veveyse de Châtel river catchment upstream from the sediment retention section are given in Table 3.

Through the city of Châtel-St-Denis, the Veveyse de Châtel has a slope of ~2%. Upstream and downstream, the slope is significantly steeper (>4%). The local slope reduction is mainly due to a sill build years ago for a mill, and currently used for small hydroelectric power plant, but the area was also formerly an active sector of sediment deposits and remobilizations, with a wider river bed than nowadays. For common floods, the river has a tendency to erode the bed (and the bank). For this reason, a series of sills have been constructed through the city in the past.

3.3 Project presentation

The project has the goal of stocking between 4,000 and 6,000 m³ of sediments in case of floods of return periods between 100 and 300 years, without impacting bed load transportation for common floods. The chosen site is characterized by an enlargement of the floodplain. Before the construction works, it was crossed by two old sills.

The spreading section is composed of two small dykes with an orifice. The orifices are dimensioned to not influence common floods, but to generate a rising of the water level in case of heavy floods. In such a case, all the area (~2,500 m²) will be flooded, and the average speed of the flow considerably slowed. This generates bed load sedimentation and an increased overflow.

3.4 Intermediate conclusion

The expected behaviour is similar to that of a river overflowing its floodplain. The advantage of the structures is to control the discharge which generates the overflow and the deposits.

Table 3. Main characteristics of the catchment area upstream from the sediment retention section.

| Surface (km²) | Elevation | | Flood discharge | | |
	Max. (masl)	Min. (masl)	30-year (m³/s)	100-year (m³/s)	300-year (m³/s)
24.1	2014	823	90	110	130

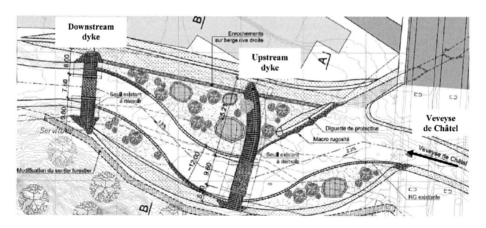

Figure 5. Topographic plan view of the project.

These structures have the same type of advantages as the sediment retention area introduced in chapter 2, but need less space and have a smaller capacity. However, common floods do tend to generate deposits which, if not removed, reduce the available storage volume. The effectiveness is therefore not as great as in the case of a retention area parallel to the river, but the solution was considered a good trade-off between moderate land consumption and effective flood protection. The cost is increased by the fact that the dikes have to be secured against overtopping and scouring by an extreme flood event, in order for them to be robust and not increase the risks once their design discharge has been reached.

4 IMPROVING SEDIMENT TRANSPORT TOWARDS THE RIVER MOUTH OF THE BAYE DE CLARENS

4.1 Introduction

The sediment contribution of the catchment area of the Baye de Clarens, a small alpine river reaching Lake Geneva in Montreux, is very important due to active erosion processes and tributaries subject to mud flows. This has led to the exploitation of several gravel pits. During the last centuries, several projects have been realized in the catchment area to stabilize the river bed and reduce the bed load transportation (sill construction, bank stabilization, etc.) and on the alluvial fan to increase the sediment transportation capacity of the river up to the mouth.

In the summers of 2005 and 2007, events with a return time period estimated below 20 years generated important but localized sediment deposits, and some damages to bridges and structures. Flooded areas were limited. The remaining hydraulic capacity of the river was locally very limited. Following these events, the decision was taken to map the natural hazard and to establish a flood protection strategy.

4.2 Catchment area and location

The main characteristics of the Baye de Clarens river catchment are given in Table 4. The Baye de Clarens has a regular and relatively steep slope in the last part up to Lake Geneva, through the city of Montreux (Fig. 6).

4.3 Project presentation

The quite regular slope of the river allows the definition of a flood protection strategy based on the increase of bed load transportation of the river instead of a retention principle. To increase the sediment capacity the replacement of five sills (between 1.0 and 3.5 m high) by ramps is foreseen. Ramps are placed in a way that their downstream end corresponds to the base of the existing sill. Such a modification requires important works in the bed and on the banks of the river, but allows an increase in the local hydraulic and sediment transport capacity of the river. It ensures an upstream migration of the sediment deposit limit (pivot point), and consequently an important reduction of the deposits upstream of the ramp.

The increased capacity of the river limits the risk in the urban area but will also increase the volume of sediment which will reach the lake. Nowadays, the mouth in the lake has to be

Table 4. Main characteristics of the catchment area.

| Surface (km²) | Elevation | | Flood discharge | | |
	Max. (masl)	Min. (masl)	30-year (m³/s)	100-year (m³/s)	300-year (m³/s)
14.3	1752	372	53	65	74

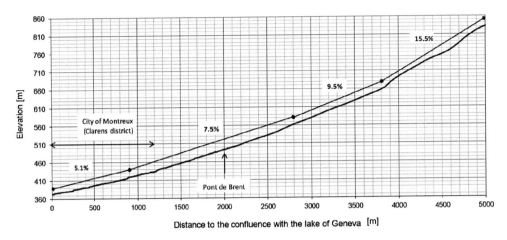

Figure 6.   Longitudinal profile of the Baye de Clarens.

dredged regularly (on average every two years) to avoid an increase of the bed level upstream of the mouth, and consequently overflows. To limit the dredging and increase the security in case of heavy floods, the size of the mouth will be multiplied by three, sharply increasing its volume. Furthermore, two local sills will be replaced by ramps to increase the energy of the flow and push the sediment into the lake.

The geometry and operation of the widened mouth has been tested and validated on a physical model by the Laboratory of Hydraulic Research (LCH) of the EPFL.

The numerical modeling of the river has shown that the planned measures described above are sufficient for an event with a return time period of about 100 years, for rarer events the bed load transportation capacity will be insufficient in different locations. For this reason, a sediment retention area is planned on the upper part of studied section.

This sediment retention area (of 4,000 m³) has to be hidden for the flood up to a hundred year return period. The design of this structure is a sill with two notches. The notches will be flowing full for a major flood; this will reduce the velocity of the flow and generate deposits. After the flood, the sediment retention area must be dredged if it is full. As an alternative, if the sediment retention area is partially full, this area can be used as a bed load supply sector.

The design calls for two small notches to be built instead of a larger one to increase the surface which can be eroded after a flood. Such a structure has really to be understood as a sediment regulation structure and not as a sediment extraction area.

4.4   Intermediate conclusion

The characteristics of the river, mainly its slope, allow considering a flood protection strategy which includes the conservation of full sediment dynamics up to a 100-year flood and the reduction of the current maintenance. Ecological benefits derive from the suppression of existing sills, while the attractiveness of the urban landscape is increased thanks to the wider mouth area.

5   CONCLUSION

During the last decades, the answer to the problems caused locally by sediments along small alpine rivers, especially on their alluvial fans, has been the construction of sediment dumps, combined with regular sills in the river to prevent downstream erosion. This solution has shown its limits in terms of robustness and generated a depletion of the ecological quality of the river, due to the lack of sediment.

Increasing environmental awareness, the recent development of sediment computing software and the growing experience with physical models and real scale projects make it possible to conceive flood protection measures which include the notion of sediment dynamics. Such an approach fits perfectly with a sustainable vision of flood protection satisfying the legal requirements for more natural rivers while also complying with the obligation to achieve robust technical solutions that will not fail under floods exceeding the design discharge, a condition that becomes increasingly more important in the face of climate change.

REFERENCES

Ghilardi T., Boillat J.-L., Schleiss A., de Montmollin G., Bovier S., 2012. "Gestion du risque d'inondation sur l'Avançon. Optimisation sur modèle physique", 12th Congress INTERPRAEVENT 2012, Grenoble, France.
GHO—Groupe de Travail pour l'Hydrologie Opérationnelle (GHO), 1999. Recommandations concernant l'estimation de la charge sédimentaire dans les torrents—Parties 1 et 2. Service hydrologique et géologique national, Berne.
Hager W.H., Schleiss A., 2009. Constructions hydrauliques—Ecoulement stationnaires. Traité de génie civil volume 15. PPUR.
Jaeggi, M., 1992, Sedimenthaushalt und Stabilität von Flussbauten, Mitteilung der Versuchsanstalt für Wasserbau, Hydrologie und Glaziologie der ETH Zürich Nr. 119.
Jaeggi, M., 1995, Flussbau, ETH Zürich, chair de constructions hydrauliques.

Swiss Competences in River Engineering and Restoration – Schleiss, Speerli & Pfammatter (Eds)
© *2014 Taylor & Francis Group, London, ISBN 978-1-138-02676-6*

Integrated flood forecasting and management system in a complex catchment area in the Alps—implementation of the MINERVE project in the Canton of Valais

J. García Hernández & A. Claude
CREALP, Sion, Switzerland

J. Paredes Arquiola
UPV, Valence, Spain

B. Roquier
HydroCosmos S.A., Vernayaz, Switzerland

J.-L. Boillat
LCH, EPFL, Lausanne, Switzerland

ABSTRACT: A complex hydrologic-hydraulic model has been developed for the Upper Rhone River basin in Switzerland. It is currently operational in the Canton of Valais for real-time flood forecasting and management, providing automatic warnings to the crisis cell of the Canton as well as proposing preventive emptying operations of dam reservoirs to reduce potential flood damage. The system is connected with a database for real-time data transfer and a website has been created to provide information for flood management, such as warning levels, hydrological forecasts at the main control points of the Rhône River and its tributaries, precipitation forecasts over the whole basin, snow cover state and reservoirs water levels. Besides, a hydrological call center has been established for supporting the crisis cell during risked event situations.

1 INTRODUCTION

1.1 *Framework*

The project of the Third Rhone Correction upstream from Lake Geneva improves the plain protection level during floods. In collaboration with this river engineering project, the MINERVE system (García Hernández et al. 2013; García Hernández 2011; Jordan 2007) aims to predict and optimize the management of flows using the hydrological forecasts and the hydropower scheme operation data. A model for hydro-meteorological forecasts coupled with a decision support system for preventive operating strategy of the hydropower plants have been developed and implemented.

1.2 *Objectives*

The objective of the operational system implemented in the Canton of Valais is to provide hydro-meteorological information to improve flood management in the Upper Rhone River basin. To achieve this task, a cluster for flood forecasting and management has been created at the Research Center on Alpine Environment (CREALP). This multidisciplinary group is operating a real-time flood forecast system that provides hydrological forecasts with the RS MINERVE tool (chapter 4.2) at the main control points of the Rhone River and its tributaries. Based on the hydrological forecasts, automatic warnings, associated with four different thresholds at each control point, are provided to the crisis cell. Finally and also based on these forecasts, preventive operations can be proposed to hydropower schemes owners to reduce as much as possible the potential flood damages.

2 UPPER RHONE BASIN

2.1 *The basin*

The Upper Rhone River basin (Fig. 1) is located in the Swiss Alps, upstream from Lake Geneva. It covers a surface of 5'524 km^2, including 658 km^2 of glaciers, and is characterized by high mountains with elevations varying from 372 to 4'634 m a.s.l.

The total length of the Rhone River, from its source at the Rhone Glacier over 2'200 m a.s.l. to the Lake of Geneva at 372 m a.s.l., is around 165 km. The average year discharge between January 1st 1980 and January 1st 2014 at Porte du Scex, outlet of the basin, was 189 m^3/s, and the highest discharge 1'358 m^3/s, measured on October 15th, 2000.

Many hydropower schemes with large reservoirs are located in the watershed, strongly influencing the hydrological regime of the river network. The reservoirs have a total storage capacity of 1'195 Mio m^3 and a total equipped discharge of more than 500 m^3/s.

2.2 *Spatial discretization of the basin*

The watershed (Fig. 1) has been divided into 245 sub-basins taking into account not only the Digital Elevation Model (DEM), but also the gauging stations and rivers junctions as well as reservoirs, hydropower plants and intakes. Each sub-basin was next discretized into different elevation bands allowing temperature adjustment. The model also takes into account the glacial cover with the last available data of swissTLM3D, being the 1347 defined elevation bands distributed into non-glacial (1027) and glacial (320). For the non-glacial bands, the average surface is 4.7 km^2 and the average of the vertical altitude difference of each band is

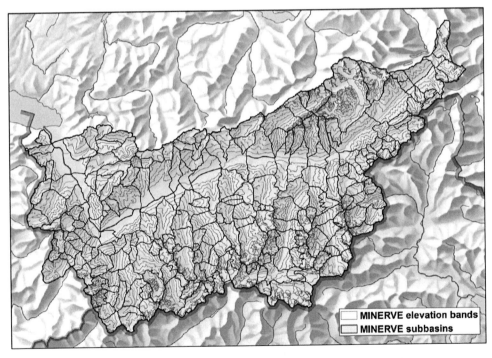

Figure 1.   Upper Rhone River basin with its 245 sub-basins and 1347 altitude bands.

361 m. For the glacial bands, the average surface is 2.1 km² and the average vertical space 327 m.

The Rhone River has been divided into 22 river reaches and the tributaries still bring 141 additional river segments to complete the hydrological model.

Finally, significant storage reservoirs (14), compensating basins (13), hydropower plants (31) and intakes (91) have been included in the model.

2.3 *Hydrological model*

Based on the spatial discretization of the basin, a semi-distributed model was developed to simulate the hydrological and hydraulic behavior of the river basin. The model includes snow-melt, glacier melt, soil infiltration, surface runoff, flood routing in rivers and reservoirs as well as hydropower plants operations.

To take into account the hydrological processes, different hydrological models to calculate rainfall-runoff processes are provided by the RS MINERVE numerical tool (chapter 4.2).

The GSM (Glacier and SnowMelt) model (Schäfli et al. 2005) calculates the snow and glacier melts with a degree-day approach. The new SOCONT (Soil Contribution) model includes a snow model, a GR3 model (Michel and Edijatno 1988; Consuegra et al. 1998) calculating infiltration and a SWMM (Storm Water Management Model) model (Metcalf and Eddy, 1971) for runoff simulation.

The model HBV (Bergström 1976, 1992) is an integrated rainfall-runoff model, which includes conceptual numerical descriptions of hydrological processes at the catchment scale. It also calculates the snow processes as well as runoff and slow and rapid infiltration flows.

The GR4 J model is a hydrological model (Perrin et al. 2003) which simulates the flow from a production and a routing stores based on two unit hydrographs.

Finally, the SAC-SMA (Sacramento Soil Moisture Accounting Model) model (Burnash et al. 1973; Burnash 1995) calculates the runoff and the baseflow distinguishing different soil zones, with rational percolation characteristics.

Based on the concept of the elevation bands per sub-basin, the model was built from upstream as presented in Figure 2. The elevation bands of each sub-basin supplies the same outlet. Then, the discharge flows through a river reach to the next junction, where contributions from other different sub-basins can be added.

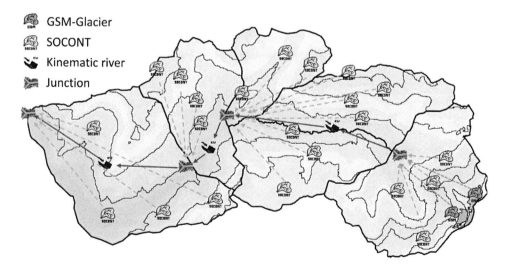

GSM-Glacier

SOCONT

Kinematic river

Junction

Figure 2. Sub-basins and elevation bands in the Grande Eau watershed (tributary in the downstream part of the Upper Rhone River basin) and their schematic representation as glacier and non-glacier bands, rivers and junctions.

Potential Evapotranspiration (PET) can be provided by the database if available. Since only precipitation and temperature values are available at this moment, different methods have been implemented in RS MINERVE for PET estimation. Turc (1955, 1961) equation as well as McGuinness and Bordne (1972) calculate PET based on current temperature and global radiation. Oudin (2004) methodology provides PET values based on current temperature and estimated extraterrestrial radiation.

The value of global radiation R_g is calculated per month (average value) depending on the latitude and longitude of the river basin and is based on 22 years monthly averaged (July 1983–June 2005). These R_g data were obtained from the NASA Langley Research Center Atmospheric Science Data Center Surface meteorological and Solar Energy (SSE) web portal supported by the NASA LaRC POWER Project (http://eosweb.larc.nasa.gov/sse/).

3 HYDRO-METEOROLOGICAL INPUTS

3.1 Meteorological data

Meteorological observed values are provided by MeteoSwiss (31stations) and SLF (52 stations) networks (Fig. 3), respectively at a ten and thirty minutes time step. The MeteoSwiss stations provide precipitation, temperature, wind (speed and direction) and relative humidity data and are located at altitudes between 380 and 2'500 meters (mean of 1'700 meters). The SLF stations provide, depending on each station, values such as precipitation, temperature, snow height, wind direction and intensity or humidity. They are mainly installed for avalanche risk evaluation, delivering a good representation of the high relief meteorological variability, at altitudes between 2'000 and 3'300 meters (mean of 2'662 meters).

For the computation of flood predictions, the meteorological forecast model COSMO-7, provided by MeteoSwiss, is available. It is driven by the global model ECMWF (European Centre for Medium-Range Weather Forecasts) and covers most of Western and Central

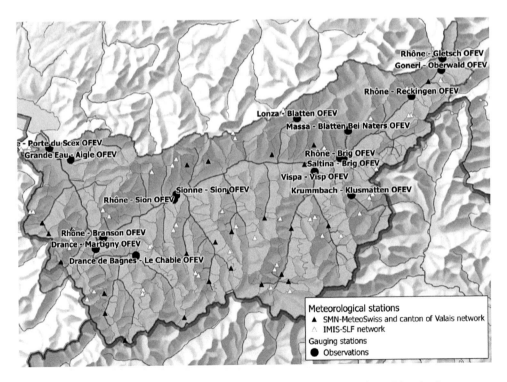

Figure 3. Hydro-meteorological data available in real-time in the Upper Rhone River basin.

90

Europe. Its spatial resolution is 6.6 km and the temporal resolution is 1 h. The horizon is 72 h and it is updated three times a day at 00:00, 06:00 and 12:00 UTC.

3.2 Hydrological data

Gauging stations are mainly operated by the Federal Office for the Environment (FOEN). Discharge series in 17 FOEN stations located in the basin are available since more than 30 years ago and are nowadays obtained in real-time (Fig. 3) each 10 minutes or 1 hour depending on the stations.

In addition, 6 cantonal stations measuring water levels are also available in real-time since 2013 in some Rhone tributaries, but they are not yet used in the operational system. Their relationship between level and discharge is being currently calculated and will allow improving the model results by providing additional control points for calibration. Furthermore, other cantonal stations will be installed in 2014 in other tributaries and will also improve the model performance, especially for the concerned lateral valleys.

3.3 Snow data

To calculate the snow cover, a processing for MODIS (Moderate Resolution Imaging Spectroradiometer) data was developed to obtain observed snow extent and its evolution. This information is later used for updating the snow cover considered in the hydrological model.

The MODIS data are produced by an imaging radiometer for Aqua and Terra satellites. Both satellites are part of the EOS program (Earth Observing System) from the NASA. The CREALP uses cryosphere data (snow and ice on land) from MODIS, which are freely available on the National Snow and Ice Data Center website. The evolution of the snow cover and land can then be observed and studied. The snow cover is calculated once per day, with an average delay of two or three days (Fig. 4). In addition, historical data since February 2000 are available and allow calculating the snow cover norm per day or per month (Fig. 5).

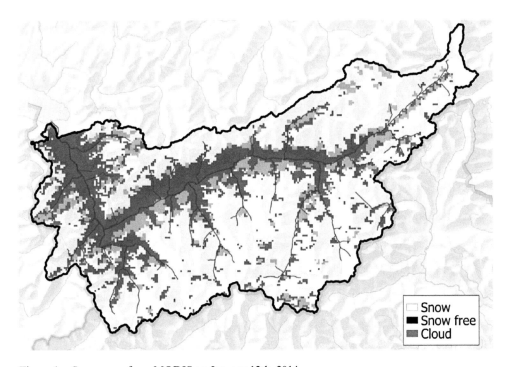

Figure 4.   Snow cover from MODIS on January 12th, 2014.

91

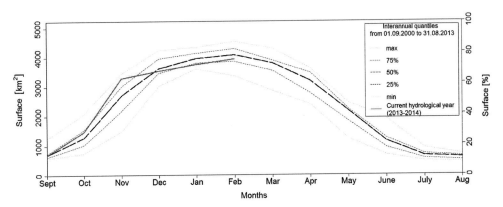

Figure 5. Quartiles 0, 25, 50, 75 and 100% with the snow cover of the Upper Rhone River basin calcu-lated with MODIS data from 01.09.2000 to 31.08.2013, as well as the current situation from September 2013 to February 2014.

4 OPERATIONAL SYSTEM

4.1 *Operational scheme*

The implemented real-time flood forecasting system (Fig. 6) is composed of: a database for hydro-meteorological data storage and real-time data transfer; a computer for automatic hydrological simulation (hindcasting with observed meteorological values and forecasting with meteorological forecasts); and a website for hydro-meteorological data information, where observations and forecasts are presented through graphics and tables.

The database stores all hydro-meteorological information that it receipts from the different providers. Afterwards, meteorological (precipitation and temperature observations and fore-casts) and hydrological (discharge measurements) data are sent to the computer dedicated to hydrological calculations. The hydrological simulations are realized with the aid of the tool RS Pilot (chapter 4.3), which manages the hydrologic-hydraulic software RS MINERVE.

The hydrological outputs (discharge forecasts at main control points) are sent to the data-base to be stored. Then, data are transmitted and published in the information website pol-hydro.ch. This one has been created with the objective of displaying all useful information for managing floods, such as warning levels, hydrological forecasts at main control points of the Rhone River and its tributaries, precipitation forecasts over the whole basin, snow cover state and reservoirs water levels, among other information.

4.2 *Hydrological/hydraulic simulations with RS MINERVE*

Routing System II (Dubois and Boillat 2000; García Hernández et al. 2007), was developed at the Laboratory of Hydraulic Constructions (LCH) at the Ecole Polytechnique Fédérale de Lausanne (EPFL). This program simulates the free surface run-off and its propagation. It models hydrological and hydraulic complex networks according to a semi-distributed conceptual scheme. In addition to particular hydrologic processes (snowmelt, glacier melt, surface and underground flow), hydraulic control elements (e.g. gates, spillways, diversions, junctions, turbines and pumps) can be also incorporated.

Built upon Routing System II, RS MINERVE (Foehn et al. 2014; García Hernández et al. 2014) was almost completely recoded to obtain a program that is both efficient and user-friendly for hydrological and hydraulic modeling.

The global analysis of a hydrologic-hydraulic network is essential for the development and optimization of an appropriate flood protection concept or the potential role of hydro-power plants. RS MINERVE (Fig. 7) allows such kind of analysis due to its simple and intuitive interface and its open structure facilitates the implementation and adaptation of

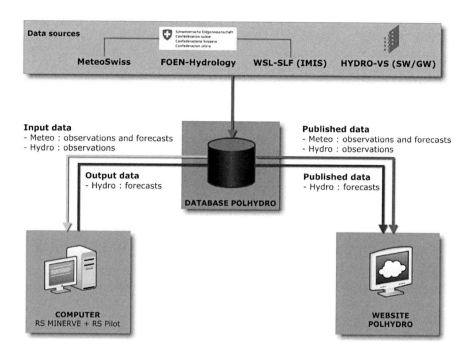

Figure 6.   Operation of the RS MINERVE system for flood forecasting in the Canton of Valais.

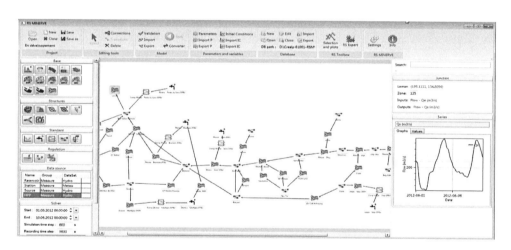

Figure 7.   RS MINERVE main interface.

new developments. The implementation of an RS Expert section, with 3 modules detailed hereafter, allows easily improving the model performance and better evaluating hydrologic and hydraulic results.

The automatic calibration module, with the SCE-UA algorithm (Duan et al., 1992, 1993, 1994), calculates the best set of hydrological parameters depending on an objective function defined by the user and becomes a key element of the software. This new module is being currently used for improving the model performance based on different criteria such as Nash Coefficient, Relative Volume Bias or Pearson Correlation.

The second module for time-slice simulations facilitates the analysis of large data sets without overloading the computer memory. It allows the evaluation of the model results performance over the whole Rhone river basin for the complete database, calculating it year by year.

The third module for scenario simulations introduces the possibility of simulating multiple weather scenarios or several sets of parameters and initial conditions to evaluate the variability and sensitivity of the model results. It is useful in flood situations for providing additional information about the situation and assessing the uncertainty of the forecast and/or the model.

Finally, data assimilation is used to improve the results obtained with RS MINERVE since the beginning of 2014. The procedure is currently manual and the assimilation is undertaken once the performance of the model is not satisfactory.

Discharge series at 17 measurements points in the basin are used for updating the model and the saturation balance is modified to achieve enhanced discharge values. Furthermore, snow cover (MODIS) is applied to adjust the snow model extent considered by the hydrological model and provide reliable flow from snow melt. Finally, the available levels of reservoirs are also used to update the model, starting each new simulation from the correct reservoir level.

4.3 RS pilot

RS Pilot was developed to perform automatic simulations with RS MINERVE. To achieve this goal, a timetable is defined beforehand, which contains the time at which the system imports the meteorological forecast, runs the model and exports the hydrological results.

To perform hydrological forecasts, two different tasks are managed by RS Pilot. On one hand, the "Control simulation" task allows running the hydrological simulation with the last meteorological observations to get the hydrological state variables, such as snow height or soil saturation. On the other hand, the "Forecast simulation" task takes into account the last meteorological forecast and the hydrological state variables obtained by the previous task.

4.4 Operational center of hydro-meteorological monitoring

A hydro-meteorological call center has been established to support the crisis cell of the Canton of Valais during critical events. A pool of operators was created and trained on the different meteorological and hydrological tools to be able to ensure the hydro-meteorological monitoring. This call center provides an important scientific and technical support for the flood risk management.

The supervision is mainly based on the data monitoring. On one hand, data such as current and previous rainfall, temperature or snow cover per sub-basin or discharge at the main control points are checked. On the other hand, meteorological forecast provided by MeteoSwiss are analyzed at the same time as the hydrological forecasts at the main control points.

The hydrological flow forecasts are computed at each sub-basin of the model. Among them, 23 control points (Fig. 8), corresponding to the gauging stations (for comparison reasons) and to sensitive locations like confluences or bridges, are presented in the website polhydro.ch.

Depending on the precipitation, MeteoSwiss proposes several danger levels from 1 (no or minor danger) to 5 (very high danger) on 23 regions over the Rhone River basin. In addition, 4 of the control points located on the Rhone River are used to indicate the same danger levels (1 to 5) depending on different discharge thresholds provided by the Canton of Valais.

Contacts with the Geneva MeteoSwiss forecasters are established when the meteorological situation deserves further information. Other contacts and information exchange with the Department for Roads, Transports and Watercourses of the Canton of Valais as well as with the Swiss Federal Office for Environment (FOEV) are also maintained when discharge values go near the second threshold (third danger level).

The website polhydro.ch provides latest hydro-meteorological information to the operators following the situation as well as to the crisis cell of the Canton of Valais for taking possible measures (structural protection measures, information to the population, village evacuation, etc). An automatic report with an overview of the current available information is also sent daily by mail to the concerned people and can be updated if required.

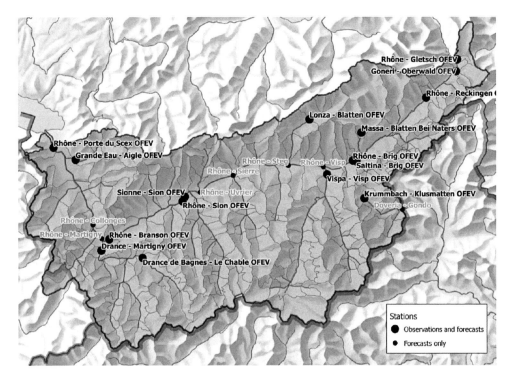

Figure 8. Hydrological control points with discharge forecasts.

5 HYDROLOGICAL FORECASTS

The MINERVE system provides 3 hydrological forecasts per day (00:00, 06:00 and 12:00 UTC) based on COSMO-7 meteorological forecasts. Precipitation and temperature are used as input of the hydrological model. Potential evapotranspiration is calculated from the Turc (1955, 1961) method and hydropower operations are estimated beforehand based on turbining scenario.

The system, operational from the beginning of 2013, provides warnings for 4 thresholds in 4 control points located on the Rhone River. An overview of the warnings performance for the first threshold during the 2013 year is presented hereafter at these 4 control points (Fig. 9). The performance of the warning system is continuously monitored and each warning is sorted as a Hit, a Miss or a False alarm. A Hit is recorded if observation and forecast exceed the threshold for a given day; if only the observation exceeds the threshold, a Miss is registered; and finally when only the forecast exceeds the threshold, a False Alarm is considered.

The results show a low performance for the control point of Brig (threshold $T1 = 250$ m³/s), due to overestimations of the snow coverage and the precipitation forecast. At control points of Sion ($T1 = 460$ m³/s) and Branson ($T1 = 470$ m³/s), results were more compelling although several false alarms were noticed due to an overestimation of the precipitation. At the outlet of the basin, Porte-du-Scex ($T1 = 700$ m³/s), results gave some miss events due to underestimated precipitation forecasts over the lower part of the basin and small saturation of the soil, as well as one false alarm due to a too high snow cover.

In 2014, the revision of the initial conditions of the hydrological model based on MODIS snow cover and discharge data from gauging stations improves hydrological forecasts and, then, the reliability of the warning system. In addition, new cantonal gauging stations provide data to recalibrate the entire model.

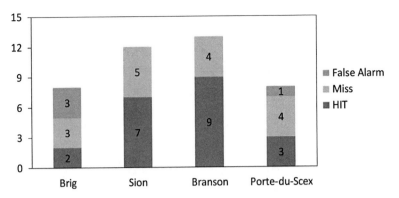

Figure 9. Hit, Miss and False Alarm results for Brig, Sion, Branson and Porte-du-Scex in 2013.

6 CONCLUSIONS

The MINERVE system is integrated in the cantonal procedure for the Rhone River flood management. It allows simulating the discharge in the river network of the Upper Rhone catchment area since it considers all hydropower plants and dams, turbine operations and water releases.

The objective of the system is to establish an optimal coordination of actions to reduce the risk of flooding taking into account present and future hydro-meteorological situation over the whole basin (Rhone River and its tributaries). The approach is based on exploitation and monitoring of observed data as well as on hydro-meteorological forecasts. The system provides the useful information to the cantonal decision-making crisis cell, which determines the level of danger and triggers appropriate protective measures or interventions.

With a constant concern for consolidation and improvement, the MINERVE system continuously modernizes its operational environment at the same time that hydrological and hydraulic issues are studied and carried out for improving the performance of the system and decreasing its uncertainty.

REFERENCES

Bergström, S. 1976. Development and application of a conceptual runoff model for Scandinavian catchments. Ph.D. Thesis. SMHI Reports RHO No. 7, Norrköping.
Bergström, S. 1992. The HBV model—its structure and applications. SMHI Reports RH, No. 4, Norrköping.
Burnash, R.J.C., Ferral, R.L., and McGuire, R.A. 1973. A generalized streamflow simulation system—Conceptual modelling for digital computers. US Department of Commerce, National Weather Service and State of California, Department of Water Resources, p 204, 1973.
Burnash, R.J.C. (1995). The NWS River Forecast System—catchment modeling. In: Singh, V.P. (Ed.). Computer Models of Watershed Hydrology, 311–366.
Consuegra D., Niggli M. and Musy A. 1998. Concepts méthodologiques pour le calcul des crues. Application au bassin versant supérieur du Rhône. Wasser, Energie, Luft—eau, énergie, air, Heft 9/10, 223–231.
Duan, Q., Gupta, V.K. and Sorooshian, S. 1993. A shuffled complex evolution approach for effective and efficient global minimization. Journal of Optimization Theory and Applications, Vol. 76, 501–521.
Duan, Q., Sorooshian, S. and Gupta, V. 1992. Effective and Efficient Global Optimization for Conceptual Rainfall-Runoff Models. Water Resources Management, Vol. 28, 1015–1031.
Duan, Q., Sorooshian, S. and Gupta, V.K. 1994. Optimal use of SCE-UA global optimization method for calibrating watershed models. Journal of Hydrology, Vol. 158, 265–284.
Dubois, J. and Boillat, J.-L. 2000. Routing System—Modélisation du routage des crues dans des systèmes hydrauliques à surface libre. Communication 9 du Laboratoire de Constructions Hydrauliques, Ed. A. Schleiss, Lausanne.

Foehn, A., García Hernández, J., Claude, A., Roquier, B., Paredes Arquiola, J. and Boillat, J.-L. 2014. RS MINERVE—User's manual v1.9. *RS MINERVE Group*, Switzerland.

García Hernández, J. 2011. Flood management in a complex river basin with a real-time decision support system based on hydrological forecasts. *PhD Thesis N°5093, Ecole Polytechnique Fédérale de Lausanne, EPFL*, Switzerland, and Communication 48 du Laboratoire de Constructions Hydrauliques, Ed. A. Schleiss, EPFL, Lausanne.

García Hernández, J., Boillat, J.-L., Feller, I. & Schleiss A.J. 2013. *Présent et futur des prévisions hydrologiques pour la gestion des crues. Le cas du Rhône alpin.* Mémoire de la Société vaudoise des Sciences naturelles 25: 55–70. ISSN 0037-9611.

García Hernández, J., Jordan, F., Dubois, J. and Boillat, J.-L. 2007. Routing System II: Flow modelling in hydraulic systems. *Communication 32 du Laboratoire de Constructions Hydrauliques*, Ed. A. Schleiss, EPFL, Lausanne.

García Hernández, J., Paredes Arquiola, J., Foehn, A., Claude, A., Roquier, B. and Boillat, J.-L. 2014. RS MINERVE—Technical manual v 1.6. *RS MINERVE Group*, Switzerland.

Jordan, F. 2007. Modèle de prévision et de gestion des crues—optimisation des opérations des aménagements hydroélectriques à accumulation pour la réduction des débits de crue. *PhD Thesis N°3711, Ecole Polytechnique Fédérale de Lausanne, EPFL*, Switzerland,; and Communication 29 du Laboratoire de Constructions Hydrauliques, Ed. A. Schleiss, EPFL, Lausanne.

McGuinness, J.L. and Bordne, E.F. 1972. A comparison of lysimeter-derived potential evapotranspiration with computed values. *Technical Bulletin 1452, Agricultural Research Service*, U.S. Department of Agriculture, Washington D.C., 71 pp.

Metcalf and Eddy, Inc., University of Florida, and Water Resources Engineers, Inc. 1971. Storm Water Management Model, Vol. I. Final Report, 11024DOC07/71 (NTIS PB-203289), U.S. EPA, Washington, DC, 20460.

Michel, C. and Edijatno 1988. Réflexion sur la mise au point d'un modèle pluie-débit simplifié sur plusieurs bassins versants représentatifs et expérimentaux. CEMAGREF Antony.

Oudin, L. 2004. Recherche d'un modèle d'évapotranspiration potentielle pertinent comme entrée d'un modèle pluie-débit global. *Thèse, Ecole Nationale du Génie Rural, des Eaux et des Forêts*, Paris.

Perrin, C., Michel, C. and Andréassian, V. 2003. Improvement of a parsimonious model for streamflow simulation. Journal of Hydrology 279, 275–289.

Schäfli, B., Hingray, B., Niggli, M. and Musy, A. 2005. A conceptual glacio-hydrological model for high mountainous catchments. Hydrology and Earth System Sciences Discussions 2, 73–117.

Turc, L. 1955. Le bilan de l'eau des sols. Relations entre les precipitations, l'evaporation et l'ecoulement. *Ann. Agro.* 6, 5–152, INRA.

Turc, L. 1961. Evaluation des besoins en eau d'irrigation, formule climatique simplifiée et mise à jour. *Ann. Agro.* 12: 13–49, INRA.

Swiss Competences in River Engineering and Restoration – Schleiss, Speerli & Pfammatter (Eds)
© 2014 Taylor & Francis Group, London, ISBN 978-1-138-02676-6

Flood protection and river restoration in the urban catchment basin of Cassarate river: An opportunity to restore public areas along an urban watercourse running through the city of Lugano

L. Filippini
Ufficio dei corsi d'acqua, Dipartimento del Territorio, Bellinzona, Repubblica e Cantone Ticino, Switzerland

S.A. Ambroise
Officina del paesaggio, Lugano, Switzerland

S. Peduzzi
Ufficio dei corsi d'acqua, Bellinzona and Section des sciences de la terre et de l'environnement, Institut Forel—Centre de biologie alpine Piora, Université de Genève, Switzerland

ABSTRACT: In the next decades the city of Lugano will develop in a territory which will coincide with the catchment basin of the Cassarate river. This river course will be considered as the load-bearing axis of the "new city" and has the potential to become a new green and recreational axis running through the metropolitan area connecting neighborhoods. To meet these goals a large multi-purpose project on Cassarate river is ongoing. Indeed, according to Cassarate hazard zone map, some metropolitan areas still do not meet flood protection requirements. Moreover the areas surrounding Cassarate river are ecologically degraded and are only accessible to a limited extent for recreational use. The Cassarate restoration project will thus combine hydraulic measures with ecological and recreational restoration measures. Restored areas, surrounding Cassarate river, will thus allow several recreational and public activities. They have the potential to become high quality public spaces for local people as well as for visitors and tourists. They will also provide natural areas within the city and promote ecological interconnections for fishes and other organisms along the river course.

1 INTRODUCTION

The Cassarate river, flowing down from Valcolla and Valle Capriasca, runs through the city of Lugano before flowing into the lake of Lugano between the Ciani Park and the city's Port. The river management project, which is designed primarily to meet safety needs, became, during development, an opportunity to enhance the urban environment. It is also an opportunity to strengthen the role of Cassarate river as a link between public areas and various functional elements, all of which are of vital importance for the city. The combination of two major urban projects, the New Cornaredo District (NQC) and enhanced mobility in the urban area, the Hub Road System Plan, PVP, has made it possible to obtain important synergies. The political situation and an aggregation process which has just been completed, give this flood protection project, a symbolic importance and a new function as a link between the new neighborhoods.

2 SITUATION

The Cassarate river is 18.3 km long and its catchment area is limited by channelization in the urbanized stretch. Its average gradient is 7%, with over 25% in the upper section, whilst the final

5 km within the built-up area, are flat with gradients of approximately 1%. The catchment area, of which almost 70% is now within the municipality of Lugano, covers an area of 73.9 km² and generates an average flow rate of 2.5 m³/s. Its sources are at 2050 m above sea level, in the upper section of Val Colla, nearby Monte Gazzirola and Passo del San Lucio. The river flows through the municipalities of Capriasca, Canobbio and Lugano (Fig. 1). It enters the lake of Lugano not far from the city centre, in the *Foce* area, near Ciani Park and Lido di Lugano.

An analysis of the available hydrographical data from Lugano's hydrometric station, Cassarate—Pregassona (2321), run by the Federal Office for the Environment FOEN (http://www.hydrodaten.admin.ch/it/2321.html) illustrates the torrential nature of the Cassarate river, which can have considerable variations in the flow rate (Salvetti, 2011). Normal situations with low water levels, and discharges lower than or equal to 1 m³/s, are interspersed with high water levels which may reach 180 to 200 m³/s (HQ_{30} = 150 m³/s, HQ_{100} = 220 m³/s); during the flooding of 1951 and the most recent events in 2001 (with a HQ = 160–180 m³/s), 2002 and 2008, the river overtopped its banks in the city and erosion occurred in the Piano della Stampa and at the mouth of the river, in the *Foce* area.

Studies of natural hazards show that currently there is a significant lack of protection in sensitive, highly urbanized areas in regions bordering the river; infrastructures whose func-

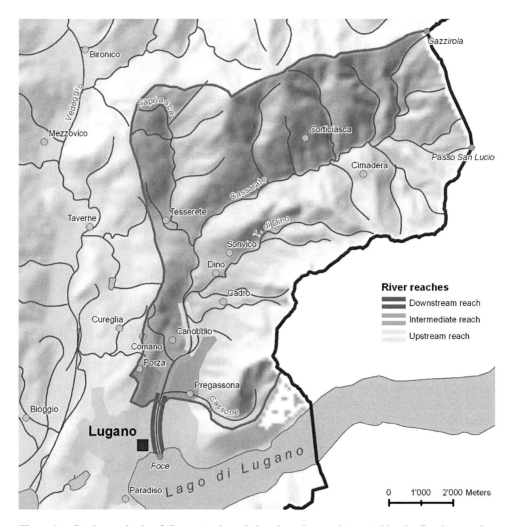

Figure 1.   Catchment basin of Cassarate river. Colored reaches are interested by the flood protection and restoration project (figure concept: L. Krebs & A. Salvetti, Ufficio dei corsi d'acqua).

100

tionality may be compromised include communication routes and public and private buildings. The Hazard Zone Map, PZP, clarifies the situation (Fig. 2, left).

The river Cassarate was during a long time a dividing line between the city of Lugano and the neighboring municipalities. After the conclusion of the recent process of urban aggregation which profoundly transformed the city's territorial and administrative structure, the river has ceased to be an administrative boundary and has taken on a central role.

The Cassarate watercourse connects essential structures and functions. For the city of Lugano, the River Cassarate nowadays represents a significant structuring axis which comprises and connects an important network of green areas: along its left bank a "green belt"

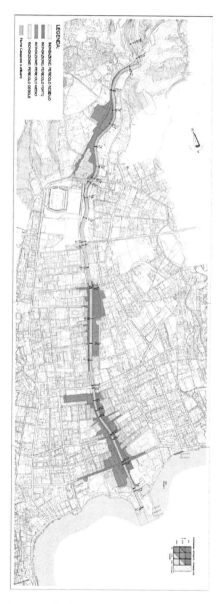 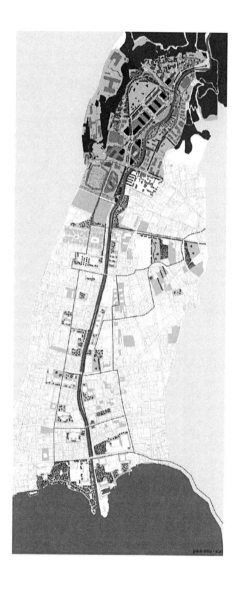

Figure 2. Left: Cassarate hazard zone map of river reach Ponte di Valle (km 3.7)—Foce (river mouth) (from: Tunesi Ingegneria SA, 3rd November 2006). Right: the new "green" axis developing along the urban watercourse of Cassarate, the continuous red line indicate the pedestrian and bicycle path (from: Nuovo Quartiere Cornaredo—Lugano, www.cornaredo.ch).

is planned, which allows "slow mobility" activities like bicycle and pedestrian path (Fig. 2, right).

The banks of Cassarate River act as a link between different functions playing a key role: from the north, to the confluence of the Dino stream where there is the Piano della Stampa industrial area and where redevelopment is planned. Proceeding orographically, The New Cornaredo District (NQC, see Fig. 3) being built, takes in the East portal of the Vedeggio—Cassarate road tunnel, providing new road access to the city with the Hub Road System Plan, PVP. This is then followed by sports infrastructure, the stadium and ice rink; the university campus which is being developed astride the two banks, the ex-slaughterhouse area, Campo Marzio and the exhibition pavilion; after the Viale Cattaneo bridge, in the *Foce* area, the Studio Foce cultural centre, the Lido, the High School in the Ciani Park and the city's Port.

Figure 3. An aerial view of the New Cornaredo District and the Cassarate watercourse marked in green (from: studio LAND and Nuovo Quartiere Cornaredo—Lugano, www.cornaredo.ch).

3 THE PROJECT AND ITS MEASURES

The course of the river is the object of a major river management project pursuing goals which fall into three distinct categories: safety, environment and public access and usage of the river banks. The project leader is the Dipartimento del territorio Cantone Ticino which in 2000 initiated a preliminary study, leading, in 2006, to the drafting of the outline project. The project affects the final reach of the river, from the Piano della Stampa down to the mouth, for a total length of 5.3 km. During the first phase of the project, the priority issues regarding the natural hazards and safety were addressed by the Tunesi Ingegneria SA project.

In 2004. with regard to managing and upgrading specifically the river mouth area, the city of Lugano organized a contest which was won by the landscape architect Sophie Agata Ambroise who—with an interdisciplinary team—drafted the final project, delivered in 2009. The Canton who manages the project divided it into three parts: an initial section, upstream, extending from the "Piano della Stampa" to "Ponte di Valle" in yellow in Figure 1, a second intermediate reach from "Ponte di Valle" to the Cassone stream in orange in Figure 1; the third reach, downstream, is the one running through the most intensely urbanized area, starting at the Cassone stream and flowing down to the mouth (in red in Fig. 1).

3.1 *Upstream reach*

A flood protection project was undertaken in the upper part of the metropolitan area of Cassarate to protect industrial and manufacture areas from riverbanks erosion which occurred during the 2001 and 2008 floods (in yellow in Fig. 1). These measures were taken using blocks of granite and natural engineering techniques. Meanwhile the channel bed has been structured with block-ramps to favor fish migration. Along the riverbanks a path for joggers and cyclists has been created. The successful upriver restoration in the Stampa area is already an example of a successful gain of valuable public spaces in a suburban area (see Fig. 4).

In this stretch the embankments (levees) have been consolidated and the banks upgraded for a total for a total of 1.5 km. The works on this first reach were finished at the end of summer 2013.

3.2 *Intermediate reach*

In this river stretch (in orange in Fig. 1), the watercourse project does not require particular alterations to the river bed and the levees. It is planned to carry out a local raising of the embankments on the right side from kilometer 3.000 to kilometer 2.810. With regard to river crossing points, it is planned to demolish the Ruggi bridge (kilometre 3.320). Within the framework of the NQC project and completion of the mobility works, it is planned to construct a roundabout sitting astride this section of the river, which will serve as a junction at the Vedeggio-Cassarate tunnel exit and to create a public area, a sort of piazza, directly facing the river. The banks of the river will also be provided with new tracks to encourage "slow mobility".

3.3 *Downstream reach*

The downstream reach (in red in Fig. 1) running through the most densely urbanized parts of the city into which the Cassone stream flows, will be will be upgraded regarding flood safety by redefining the discharge section; it is planned to lower the riverbed and, where necessary, to raise the crest of protection walls.

The new sections of the river have been dimensioned taking into account the relevant water flow: a 100-years flood ($HQ_{100} = 220$ m³/s) and an extreme flood of EHQ 310 m³/s.

Alongside hydraulic and safety requirements, the landscape—environmental preliminary project commissioned to *studio Land* (Landscape, Architecture, Nature and Development)

Figure 4. Views of the Piano della Stampa reach in 2001 before (left) and in 2013 after (right) the intervention. (Photos: Stefano Castelli—Dionea SA and Mauro Marazzi—Ufficio dei corsi d'acqua).

addresses issues of mobility and usability of spaces within the urban context in synergy with the current major roadway projects underway.

In particular the planned interventions are:

- Revitalisation and structuring of the river bed and the left bank by eliminating the artificial paved riverbed along a section of approximately 0.9 km from the Cassone stream to Ponte Madonetta (from km 2.100 to km 1.264).
- Lowering the river bed by 60 cm from Ponte Madonetta (from km 1.264 to km 0.253).
- Raising the flood retaining wall by 50 cm between 1.058 and km 0.840 (in the Università della Svizzera italiana area).
- Partial integration in the flow sections of elements to encourage "slow mobility" and enhanced usage, upgrading and widening the existing paths.

The final project for this reach will provide greater details regarding this solution with an interdisciplinary planning team starting in the second half of 2014.

With regard to crossing points, the bridge in via Fola (kilometer 1.791) has recently been replaced; the related works were completed in 2010; it is also planned to replace the pedestrian bridge in the Viganello scuole medie area (kilometer 1.394) and the one bridge in the University—Hospital area (kilometer 0.466).

3.4 Foce area

La Foce (literally the river mouth) is undergoing a thorough restoration project (see Fig. 1, tehe green dot). Major aim of the *Foce*-project was to restore Cassarate levees built in the past as vertical stonewalls, by restoring natural and vegetated riverbanks on the right side and a riprap (an array of blocks) on left side. On both sides public access and contact to water will be significantly improved (see Fig. 5).

To manage and upgrade the Cassarate River mouth area an architectural contest was organized. The project chosen by the jury is now in the final stages of completion and will be

Figure 5. A view of the river mouth *La Foce* in 2011 before (above) and in 2013 during the intervention (bottom). (Photos: Daniel Vass and Sandro Peduzzi).

inaugurated in the course of 2014. The winning concept is a project involving public space and urban parkland on both side of the river mouth. On the right side (Ciani Park) a natural bank has been restored using natural engineering techniques, recreating a floodable area and alluvial deposits with spontaneous vegetation. The inclusion of a protruding wooden walkway allows for controlled access to the bank and makes it possible to again enjoy water visual contact and a new panoramic viewpoint overlooking the lake and Monte San Salvatore. On the left side a stepped slope made of rectangular blocks of stone from the Riviera quarries is under construction. At the foot of the existing stone wall and of the new stepped slope, a submersible passageway in quarry stone blocks makes it possible to access the riverbed. Access is also possible from a new stairway upstream. Between the left bank and the port, three new terraces of river stones of varying size will make it possible to gradually access the water and to carry out sports, cultural and leisure activities.

The new pedestrian bridge allows to cross the river at the same level from the two banks. In case of flooding, a raising device next to the left-hand abutment makes it possible to guarantee safety, even in situations of extreme structural stress. In determining the levels of each infrastructure and improvement measures, high water levels and a range of lake levels were taken into consideration, considering critical scenarios.

4 ORGANISATION AND PARTICIPATION

The project to solve the Cassarate deficit situation was born through the need to keep the Lugano built-up area safe against flooding and river erosion. The project has been promoted by the Dipartimento del Territorio up to the final project stage, as the basis for the approval and funding procedure. Construction has thus been assigned to a local body, Consorzio Valle del Cassarate for the upstream stretch or Comune di Lugano for the Foce part. During project development, urban, environmental, access and usage functions along the water course have been integrated.

Difficulties during project approval and funding of the *Foce* project, by means of a popular referendum on the funding credit, promoted by the city, highlighted how citizens and interest groups are acutely aware of this kind of urban projects (Jankovsky, 2011). Thus an appropriate project organization, starting from its very early development stages, is necessary to allow the participation of all stakeholders (client, local bodies and services) to the decision making process.

5 COSTS

The total cost of the Cassarate and the river mouth project within Lugano urban area is estimated CHF 28 millions, of which 6.5 for urban functions, specifically "slow mobility" and opening up the river mouth to leisure activities.

REFERENCES

Filippini L. 2001. Un progetto urbano per lo spazio fluviale, La politica di gestione delle rive fluviali in Svizzera. Lugano: *Archi* 1/2011: 48–57.
Jankovsky, P. 2011. Die Mauer des Austosses darf fallen. *Neue Zürcher Zeitung*, 11.06.2011.
Koeppel Mouzihno L. & Peduzzi S (ed.). 2011. *Ticino: « Chiare, fresche e dolci acque »*. Mittteilungsblatt Nr. 1 Ingenieurbiologie (ISSN 1422-008).
Mariotta, S. 2011. *Il bacino del Cassarate 1880–2000, 120 anni di interventi forestali per la sicurezza del territorio*. Lugano: Edizioni universitarie della Svizzera italiana.
Peduzzi, S. & Filippini, L. 2012. Planification de la revitalization à l'échélon cantonal, entre les visions de la Confédération et celles des cantons. *Journal romand de l'environnement, Bulletin de l'ARPEA* 251: 40–44.
Peduzzi, S., Patocchi, N., Foglia, M. & Filippini, L. 2009. Gestione integrata e riqualificazione fluviale nel cantone Ticino: interventi sul fiume Ticino da Bellinzona alla foce nel Lago Maggiore. *Riqualificazione Fluviale* 2: 140–147.
Salvetti, A. 2011. *Un fiume la città e il lago Valle del Cassarate 5.2*. Bern: In viaggio attraverso il mondo dell'acqua Escursioni idrologiche in Svizzera Atlante idrologico della Svizzera.
Sassi, E. 2011. Artificio e natura. Un progetto per il fiume Cassarate. Lugano: *Archi* 1/2011: 19–23.

Swiss Competences in River Engineering and Restoration – Schleiss, Speerli & Pfammatter (Eds)
© 2014 Taylor & Francis Group, London, ISBN 978-1-138-02676-6

Intervention and management of floods in mountain rivers and torrents in the Bernese Oberland

M. Wyss
Office of Roads and Rivers Canton Bern, Switzerland

N. Hählen
Division of Natural Hazards, Office of Forests Canton Bern, Switzerland

ABSTRACT: In the last years, the Bernese Oberland has been repeatedly subject to severe flooding. In the mountainous Oberland the destruction caused by flooding often cut off entire valleys from the outside world. Since most valleys are important tourist destinations, also economical losses got substantial. Therefore, the public pressure on an immediate reconstruction of destroyed traffic and supply infrastructure and on the improvement of flood protection was significant. A well structured process had to ensure, that solutions planned and realized under immense time pressure will not be future errors. Recently, this demand could be successfully fulfilled by means of so called Local, Solution Focused Event Analyses. They comprise of an immediate, but coordinated and comprehensive analysis of the natural hazard process as well as the preliminary design of measures to reduce the protection deficiencies. The main advantage of this approach is the early availability of knowledge about the event process as an important basis for emergency and reconstruction work. Even more important is that the affected public and local authorities gain a thorough understanding of the event and subsequent protection measures.

1 INTRODUCTION

In recent years, the Bernese Oberland has been often struck by severe flooding. Main reasons were unfavorable weather conditions with either heavy short duration precipitation or rapid snow melting in combination with enduring precipitation. In addition, some flooding was triggered by changes in periglacial areas of the Alps due to global warming. On account of the very heterogeneous topography different processes occurred reaching from debris flow in steep gullies to flooding along large rivers and lakes. Side effects often were extensive erosion and extraordinary bed load transport.

As a result of the increasing number of events, the entire process from event analysis to planning of protection measures could be optimized. Optimization was necessary, since this intellectual work was governed by difficult conditions such as extreme time pressure in case of ongoing danger, anxious residents and high expectations of the public, politicians and media with respect to solutions.

2 HAZARD POTENTIAL

The Bernese Oberland is characterized by several valleys, whose rivers reach the two larger lakes of Thun and Brienz. Covering an area of 2900 km² with a total length of 5900 km of flowing waters the Bernese Oberland is the largest region within the Canton of Berne. Only three Swiss Cantons are larger. Except tourism the main economic utilization is limited to the valley bottoms. Therefore, settlements, traffic ways and other public infrastructure as well

Figure 1.   Damaged buildings and roads after the August 2005 flooding in Oey, Community of Diemtigen (Photo: A. Stocker).

as rivers are located in narrow land stripes. As a consequence, hazard mapping was enforced and favorably finished in the entire Canton by 2102. Approximately 20% of urban areas are located in well known hazard zones. At 1300 buildings—mainly in the Bernese Oberland—the allowable value of 10^{-5} p.a. for individual human death risk rate is exceeded (Arbeitsgruppe Naturgefahren 2014). In August 2005, the extraordinary flooding caused damage at 6400 buildings in the Canton of Berne resulting in a financial loss of approx. 1 billion Swiss Francs. The reconstruction of railroad lines and cantonal highways cost about 70 millions Swiss Francs (Wyss & Hählen 2006).

In case of a larger natural hazard event, the small scale topography most often results in cutting off entire valleys. During the flooding of 2005 about 60 of the Bernese Oberland's 80 communities faced limited or no accessibility due to destroyed traffic ways (Wyss & Hählen 2006). Subsequently, emergency work had to be performed with limited technical and human resources, mainly available at site. In addition to these difficulties, some communes were further stressed by supply demands as well as the need for evacuation of tourists. Especially tourist destinations asked for immediate reconstruction of traffic and public infrastructure. Quick decisions and solutions were expected. However, the risk of wrong action had to be taken into account and minimized. In light of the 70 millions Swiss Francs spent for urgent reconstruction of river courses and 150 millions invested in additional river engineering and safety projects it seems appropriate to establish a special analysis and planning process (Sect. 4).

3   EMERGENCY AND RECONSTRUCTION WORK

3.1   *Principles related to emergency and reconstruction work*

After a flood incident it is evident to obtain a minimum safety against possible subsequent events and to put main infrastructure back into service as soon as possible, at least with temporary measures. However, emergency and reconstruction work—initiated within hours or the most a few days after an event—has to be executed without negative prejudice for the final protection project. Such work has to be thoroughly planned and prioritized, no matter how urgently it should to be executed. Furthermore, emergency and reconstruction work should not lead to additional risks (Künzi & Hählen 2011). Especially after severe events occurred, it is evident to understand the cause and the process of the event before significant emergency or reconstruction work is executed. Erosion and sliding zones shall be carefully monitored without delay. Less critical work such as removing wood, debris and bed load from the watercourse or temporary protection measures against further erosion can normally be executed right away. When damage is limited and the involved processes are obvious reconstruction work can be carried out quickly as final protection measure. However, if conditions are complex and destruction extensive, it might be relevant to first come up with immediate or temporary measures to gain time for an in-depth analysis and the planning of a final project. If accessibility of upstream settlements or other important infrastructure is affected, it is of great importance to put into service any kind of roads or temporary detours even using e.g. forest roads to satisfy minimum transportation demand. The existing hazard map must be verified. Possible, future event scenarios and the required degrees of protection have to be determined as well before starting with final planning or even construction work. Because citizens, politicians and media immediately want to see construction equipment in action it is beneficial to asses each damage area according to the principles mentioned above.

3.2 Example 1: Schwarze Lütschine in Grindelwald, 2011

Emergency work performed at the Schwarze Lütschine in Grindelwald in the summer of 2011 is a typical example for handling a local and—relatively spoken—minor event. The glacier called Oberer Grindelwaldgletscher is well known to be the source of periodical flooding. The trigger typically is the outburst of subsurface water compartments. In August 2011, 13 outbursts occurred within 48 hours leading to some major flashfloods. These floods repetitively caused extensive erosion and huge bed load transport. The bed load was deposited in a section of the river bed with minimum slope at the bottom of the valley adjacent to a settlement area with a railroad station, a campground, a hostel and a gondola. The sediment deposits threatened to relocate the water flow from the river bed resulting in dangerous flooding (Hählen 2011, Oberingenieurkreis I 2012).

First, the threats lead to the evacuation of the camp ground next to the Schwarze Lütschine. After finishing the evacuation, the operational staff used 11 heavy dredging engines to excavate the river bed, even during the three days with ongoing bed load discharge. In this context, three problems had to be solved:

- Access to the river bed was limited because recreational vehicles blocked a river bank over a long distance.
- The safety of the workers during the occurrence of repetitive flashfloods.
- Location for deposition of bed load, since no landfill sites were available neither in Grindelwald, nor in its vicinity.

After the evacuation of the camp ground visitors and residents was completed, operational staff had to relocate recreational vehicles to provide access to the river bed, even using recovery vehicles. Safety of dredging workers was ensured by installing an observation post at the glacier base and a second post at the river site, where bed load was removed. Via radio, upcoming flashfloods were announced. A flow time of about 12 minutes allowed for safe relocation of the dredging personnel and equipment. According to the local emergency plan, firefighters first served at both posts. Later, members of the Civil Protection Agency took over. Thanks to the immediate and appropriate action, larger damage on buildings and infrastructure could be avoided.

By far more difficult was the question, where 70,000 m³ of removed bed load could be deposited. In the past, the general planning of landfill sites and its regional structure plan focused on the quantities of excavation material obtained at regular construction sites. Sudden extraordinary accumulation of material was not taken into consideration so far. In addition, transportation of large quantities of bed load material over longer distances is expensive and could not be afforded by the community. Also, using bed load material for concrete production or as road construction material was not possible, since the quantities of river gravel available in the Bernese Oberland exceed the demand. Therefore, alternative and local disposal routes had to be found. Some of the material was used for increasing and widening the existing flood dykes along the river. However, most of the 70,000 m³ of material had to be permanently stored at an originally temporary landfill site in Grindelwald. Because ongoing bed load transport has to be expected over an undetermined period of time, the Community of Grindelwald plans to obtain the permission for a new landfill site in the valley floor having a capacity of about 300,000 m³. In a community, where tourism is of great economical importance and where landscape protection areas were established, such a project is quite a difficult challenge.

4 LOCAL, SOLUTION FOCUSED EVENT ANALYSES

4.1 *Process applied to flood protection projects*

After severe flood disasters in mountainous regions, significant time pressure combined with complex aims require a different approach compared to minor and local problems like the glacier related flashfloods at the Schwarze Lütschine. Immediately after the extraordinary floods in most parts of the Bernese Oberland in August 2005, a new, very successful process was established within a few days, called local, solution focused event analysis (Wyss & Hählen 2006, Hählen 2006). In 2005, six local, solution focused event analyses were successfully executed under the project management of the District I in the communities of Diemtigen, Reichenbach, Lütschental, Brienz, Meiringen/Hasliberg and Guttannen.

It comprises immediate and coordinated documentation as well as analysis of each event and leads to the conceptual solution for future flood protection with the aim to reduce the existing protection deficiencies to an allowable degree. Therefore and during the start of the process, a complete team is installed at each event site including specialists in the fields of geology, geomorphology, hydraulics, civil engineering as well as landscape architecture and environment, and others if necessary.

Typically, the first step is mainly performed by geologists and morphologists—possibly even during the lasting event. They have to determine what triggered the event and what happened where (e.g. areas of failure and erosion scars, locations and quantities of bed load transport, spatial extent of flooding and destruction). Simultaneously, engineers assess the extent of damage at the river bed, its protection structures and at the traffic ways. The main advantages of an early documentation and understanding of the event are the opportunity to install—if necessary—an immediate monitoring of failure scars, sliding zones and unstable structures like bridges as well as to define appropriate emergency and reconstruction measures. Without time loss, possible future event scenarios and according protection deficiencies are identified and conceptual protection solutions are determined in a second step. Early cooperation and communication among the different team members and also involving the affected public, local authorities and the essential cantonal and federal offices allow for a fast and effective planning. Finally, during the third step, the preliminary design of the protection project is carried out in a comprehensively participative process (Sect. 4.2).

Local, solution focused event analyses performed simultaneously at all disaster sites and managed by the District I of the Cantonal Office of Roads and Rivers lead to:

- A unitary process and the application of the same analyses and planning principles at all disaster sites.
- An efficient and effective project organization by reducing interfaces.
- Time gain, especially related to the necessity to communicate with public, politicians and media.

The term 'local' refers to the area, which has to be taken into consideration. It typically comprises of one or two drainage areas only. 'Solution focused' refers to the goal, that at the end of this process the preliminary protection project including a thorough cost estimation is handed over to the community, which is responsible for the final project and its realization. The preliminary project shall:

- fully satisfy the requirements of the Hazard Prevention Division, Federal Department of Environment,
- be acceptable with respect to the later construction permit and required funding in terms of technical and environmental aspects and cost effectiveness,

–   be understood and accepted by the public and most importantly by the local residents, who are affected by the project.

Two key factors are relevant for a successful process: As mentioned above, the interdisciplinary team allows for fast and thorough action with just-in-time interaction between the team members. The second key factor is communication. Residents as well as local politicians such as municipal councils have to be involved into the analysis and planning process from the very beginning. They have to personally know the key team members in order to gain confidence as a basis for the necessary acceptance of the later protection project, which can have a significant impact on the settlement area or on residents. The early understanding of the particular natural hazard process also allows for immediate and continuous information of the public, further increasing confidence and acceptance. Beneficial for the confidence building process also are:

–   peer reviews of the results by other independent specialists, who are not involved in the project (second opinions),
–   physical hydraulic modeling of the flood protection measures including the presentation of video clips at public information meetings,
–   high credibility and communication skills of project managers and specialists.

4.2   Example 2: local, solution focused event analysis Lütschental

During the extreme rainstorm in August 2005, extensive destruction due to flooding and sediment transport occurred in the Community of Lütschental, too. The local, solution based event analysis is presented as a typical example for the event management in 2005.

The Schwarze Lütschine flows from Grindelwald to Interlaken (Fig. 2). In a short period of time, the small river changed to a wild torrent stream transporting huge quantities of debris. The village of Lütschental is located along a river section with a slope of 1%. Upstream, the Lütschine is steep with a slope of approx. 8%. About 90,000 m³ of material were removed from this steep section due to bed and side erosion (Oberingenieurkreis I 2007). The material was deposited in the almost level section of Lütschental because of its minimum transport capacity (Fig. 4). As a consequence of the filled-up river bed severe, long lasting flooding occurred. The state highway and the adjacent railroad line were destroyed at several locations, a highway bridge and a communal timber bridge collapsed, some buildings were damaged and finally, farming land was extensively covered by debris flow deposits. Furthermore, the access road as well as the railroad to the tourist destination of Grindelwald was cut off.

Immediately realizing the complexity of the event and subsequent destruction as well as the urgent need to regain accessibility for Grindelwald, the responsible engineers from

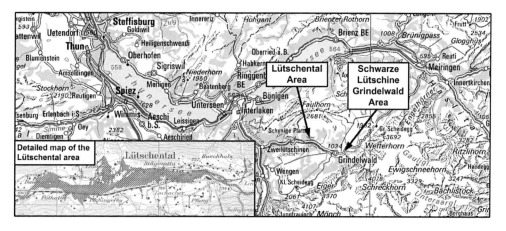

Figure 2.   Overall and detailed map of Lütschental and Grindelwald (Oberingenieurkreis I).

Figure 3.   August 2011, emergency removal of bed load (Photo: Ereigniskataster Kanton Bern).

Figure 4.   Sediment depositions and destroyed infrastructure, Lütschental 2005 (Photo: E. Gertsch).

District I decided to apply a local, solution based event analysis. Even during the ongoing flood, the team of required specialists was set up, the event documentation and analysis initiated. The chief engineer joined the regional emergency command unit for optimum coordination of all action including communication. A few hours after the process was started, the geologists and morphologists assessed a significant probability of additional side erosion and slope instabilities. Therefore, a monitoring system was installed before excavation work could be started in order to control the flooding. Knowing that reconstruction of the railroad line would take several months, a local, narrow road above the valley bottom was used as a detour route allowing for one way traffic with limited vehicle weight only. Within one month, the cantonal highway was temporary reconstructed by also installing an approx. 50 m long temporary steel bridge. Already in 2005 a regional lack of landfill sites had to be faced (Sect. 3.2). In agreement with the Cantonal Offices for Spatial Planning, for Environment and for Agriculture it was decided to primarily deposit the 90,000 m³ of bead load material in situ by locally reshaping the left valley flank, but also to use some material for the construction of a rock-trap levee to protect exposed buildings. The advantages of these measures were minimum transport and disposal costs.

Nevertheless, the total costs for all the necessary emergency work reached 7 million Swiss Francs (Oberingenieurkreis I 2007).

After completing the preliminary event analysis and emergency work, several conceptual alternatives of a flood protection project including possible relocation of the cantonal highway bridge were developed and compared. The comparison mainly comprised of assessing criteria like achievement of the required degree of safety, hydraulic system robustness, impacts on the environment and agriculture as well as cost-effectiveness. A work group of local residents, members of the municipal council Lütschental and representatives of Grindelwald as an important tourist destination closely accompanied the entire process from the beginning of the event analysis to the completion of the final design. The work group acted as a sparring partner as well as a connection link between the project team and the population. It could give inputs and formulate requests. Residents and media were informed several times, especially during the analysis and the preliminary design phase.

The comparison of the preliminary design alternatives showed, that the installation of a sedimentation management area including lowered flood plains along the river bed was the best solution. Its main advantages were cost effectiveness and minimum impact on agriculture. The highway alignment was altered to obtain a more perpendicular river crossing and to avoid the outside bank as well as sedimentation zones. The slope change of the river bed was relocated downstream of the river crossing. After carrying out a design competition the new bridge was opened to traffic in 2008. The flood protection project with costs of 8.5 million Swiss Francs was realized in stages and fully completed in 2012. The following factors led to the successful completion of this project:

– All required planning specialists were involved into the process at an early stage allowing for a straight forward and coordinated planning.
– The involvement of all essential cantonal and federal offices from the beginning was contributing to the effectiveness of the process.
– An in-partnership and equal relation with the municipal council and the population of Lütschental as well as an intensive and active communication with the public and media about the project progress.
– Direct participation of the affected population in the form of a working group, whose members represented the relevant interests of the village and the tourist destination.
– High credibility and outstanding communicative and social skills of the project managers and the representatives of the planning team.

These factors allowed for a mutual understanding between project leaders and population, which lead to a cost-effective protection project facing minimum opposition and a short planning and realization time.

5 PERSPECTIVES AND CONCLUSIONS

Due to the ongoing climate change and its subsequent impact on the alpine landscape at higher altitudes such as reduction of glacier covered areas and retreat of permafrost, a progressive increase of bed load transport has to be expected in the future. First examples are the debris flow incidents along the Grimsel Pass road (Tobler et al. 2014), in the Gasterntal above Kandersteg (Andres et al. 2012) or at the lower and upper Grindelwaldgletscher. This increase will most likely result in a demand for additional flood protection to an extent that can not be foreseen today. However, the existing natural hazard maps fortunately allow for land planning measures by preventing or restricting new settlement development in known hazard zones and therefore limit further increase of the damage potential. A major challenge remains unsolved: sufficient landfill capacity. Since the accumulation of large quantities of bed load material due to flood events can not be anticipated neither with respect to time nor with respect to location and quantity, regional landfill site planning is hardly an adequate instrument. Alternative disposal concepts have to be developed and possibly enforced. Concepts must consider options like providing more space along torrents and rivers for natural load bed deposition, abandoning usable land even in tourist destinations (agricultural and recreational land as well as proposed areas suitable for building) and the in situ deposition of material after events by remodeling landscape.

The local, solution focused analyses provide an excellent and effective instrument to manage future natural hazard events and subsequent problems like handling huge quantities of bed load material at site. They are successful, if technical challenges as well as communication and participation demands are equally approached.

REFERENCES

Andres N., Badoux A., Hilker N. & Hegg Chr. 2012. Unwetterschäden in der Schweiz im Jahre 2011. Rutschungen, Murgänge, Hochwasser und Sturzereignisse. *Wasser Energie Luft*—104. Jahrgang 2012, Heft 1: 41–49.

Arbeitsgruppe Naturgefahren des Kantons Bern (ed.) 2014. Naturgefahrenkarten Kanton Bern. Informationen zu Gefahrenstufen und Wirkungsräumen von Naturgefahren.

Hählen N. 2006. Partizipation der Bevölkerung bei der Erarbeitung eines Wasserbauplans nach dem Unwetter 2005. *Ingenieurbiologie Nr. 3/2006*: 21–25.

Hählen N. 2011. Oberer Grindelwaldgletscher. Wasserwellen und Geschiebetrieb. 25. Ereignisbericht und erste Grobanalyse.

Künzi R. & Hählen Nils. 2011. Notarbeiten während und nach Hochwasser-Ereignissen. *KOHS-Weiterbildungskurs "Gefahrengrundlagen und Hochwasserbewältigung" 3. Staffel 2011–2013*.

Oberingenieurkreis I (ed.) 2007. Lokale, Lösungsorientierte Ereignisanalyse (LLE) Lütschine. Mätzener & Wyss Bauingenieure AG, Jäggi Flussbau und Flussmorphologie, Geotest AG, Naef Hydrologie, Lehmann Hydrologie-Wasserbau, Bettschen + Blumer AG.

Oberingenieurkreis I (ed.) 2012. LLE Oberer Grindelwaldgletscher. Bericht. Hunziker Gefahrenmanagement, Flussbau AG, geo7 AG, Kinaris.

Tobler D., Kull I. & Hählen N. 2014. Hazard Management in a Debris Flow Affected Area—Spreitgraben, Switzerland. *Geografiska Annaler* (publ. in prep.).

Wyss M. & Hählen N. 2006. Unwetter 2005 im Berner Oberland—ein Überblick. FAN-Herbstkurs 2006. Meiringen. "Unwetter 2005: Lehren für das Risikomanagement".

Conference papers

Lyssbach diversion tunnel inlet near Lyss (BE), Switzerland. Photo: Esther Höck, ETH Zürich.

Swiss Competences in River Engineering and Restoration – Schleiss, Speerli & Pfammatter (Eds)
© 2014 Taylor & Francis Group, London, ISBN 978-1-138-02676-6

Driftwood retention in pre-alpine rivers

L. Schmocker
Basler & Hofmann AG, Esslingen, Switzerland

R. Hunziker
Hunziker, Zarn & Partner, Aarau, Switzerland

U. Müller
IM Ingegneria Maggia SA, Locarno, Switzerland

V. Weitbrecht
Laboratory of Hydraulics, Hydrology and Glaciology (VAW), ETH Zurich, Zürich, Switzerland

ABSTRACT: Recent flood events in Switzerland highlighted that driftwood related problems are not limited to forested mountain regions, as driftwood may reach densely populated areas under high flood discharges. Compared to mountain torrents, the damage potential may increase significantly once driftwood reaches pre-alpine rivers and larger cities. Numerous river crossing structures like bridges and weirs are prone to driftwood blocking so that the resulting backwater rise may overtop the flood levees. Prevalent retention structure as nets or racks spanned across the river are often no solution for larger rivers, because of the excessive structural loading or as the bed-load transport is prevented. Further, the driftwood potential of large catchment areas is difficult to establish and common empirical formulas for mountain torrents are not applicable to lowland rivers. Future risk analysis and engineering measures must therefore be adapted to the challenges of pre-alpine rivers. The present work summarizes techniques to predict the driftwood potential and shows two examples of driftwood retention schemes, optimized with physical scale model experiments.

1 INTRODUCTION

1.1 *Challenge*

Driftwood transport during the flood event in August 2005 in Switzerland led to a considerable increase of the flood hazard, as various cross sections of weirs and bridges got blocked (Bezzola and Hegg 2008, VAW 2008, Waldner et al. 2010). Driftwood related problems were not only limited to small mountain torrents but occurred as well in larger pre-alpine rivers. Although the discharge capacity was sufficient, driftwood blocking results in most cases in a backwater rise and a consequent overtopping of the flood levees. Various research projects and engineering measures to prevent driftwood problems were carried out since the 2005 flood at the Laboratory of Hydraulics, Hydrology and Glaciology (VAW). This paper presents selected findings obtained during the past years.

1.2 *Problem solving approach*

Project planning to solve driftwood related problems for a given catchment includes various steps. The following questions should be answered during the planning of river engineering measures: (1) How much driftwood has to be expected?; (2) What are the driftwood characteristics?; (3) Which river crossing structures are prone to driftwood blocking?; (4) Where is the optimal location for a driftwood retention structure, if needed?; and (5) Which retention structure is suitable? The following chapters give various suggestions on how to answer these questions in relation to pre-alpine rivers. Further, two recently developed driftwood retention concepts at the River Kleine Emme and the River Sihl are presented.

2 DRIFTWOOD POTENTIAL AND DRIFTWOOD TRANSPORT DIAGRAM

2.1 Overview

For a specific catchment area, the potential driftwood volume depends mainly on the volume of available driftwood and the characteristics of the flood event. The volume of available driftwood consists of dead wood or in-stream wood already distributed in the riverbed plus the fresh wood entrained during the flood (Bezzola and Hegg 2008). Trees may fall into the river as a result of changes in channel morphology and due to wind, ice loads or reduced stability due to mature age (Keller and Swanson 1979). During a flood event, fresh wood may be entrained due to bank erosion, bank undercutting, slope failures, landslides, or debris flows. The available fresh wood in river-adjacent areas depends on the wood stock, the stand density index, forest productivity and maintenance, mortality, bark-beetle or lumbering. The flood discharge determines the degree of bank erosion and strong rainfall may affect the soil saturation and consequently trigger landslides.

Driftwood includes branches, leaves, logs and rootstocks and may further consist of anthropogenic wood, e.g. wood from sawmills or wood from bridge constructions and river training structures being destroyed during a flood event. Consequently driftwood may exhibit various sizes and characteristics directly affecting its damage potential. Both the driftwood volume and its characteristics depend consequently on numerous factors and their determination for a given flood event is challenging. A certain scatter must be accepted and estimations may differ by a factor of 2 or more from the actual driftwood volume.

2.2 Empirical formulas

Various empirical formulae to estimate the transported driftwood during a flood are available. Rickenmann (1997) evaluated floods in Switzerland, Japan, and the USA presenting two formulae to estimate the effective driftwood volume depending on the catchment area and the water volume during a flood event. The data indicate a comparatively high scatter because both formulae do neither account for the catchment area characteristics nor for the flood return period.

Assuming that only the forested area adds to the transported driftwood volume during a flood, Rickenmann (1997) further estimated the potential driftwood volume using only the forested part of the catchment (for catchments < 100 km²) and the forested river length (for river lengths < 20 km). The data indicate again a high scatter with these for large catchment areas do not follow the overall trend. Uchiogi et al. (1996) evaluated various flood events in Japan and presented an empirical relation between the transported driftwood and the transported sediment volume.

The data for these formulae often originated from mountain torrents with comparatively small catchment areas so that the application to larger catchment areas is therefore critical. A comparison of the driftwood volume determined at several rivers after the 2005 flood event with the empirical formulas indicates a large scatter (Schmocker and Weitbrecht 2013). For most pre-alpine rivers, the driftwood potential is underestimated considerably when using the empirical formulae derived from mountain torrents. For pre-alpine rivers, the existing empirical formulae should therefore not be applied.

2.3 Evaluation of catchment area

A detailed analysis of the catchment area is in most cases inevitable, especially for large catchments, where various tributaries may add to the driftwood volume and the forest vegetation changes along the river course. Although various investigations exist in which the wood stock along a certain river was determined (Gregory et al. 1993, Piégay and Gurnell 1997, Brassel and Brändli 1999, WSL 2006, Böhl and Brändli 2007 or Lagasse et al. 2010), the variation between different catchment areas can be enormous. Further, the actual driftwood that may be entrained during a flood depends highly on the existing streambank protection and the channel morphology. A detailed site inspection may results in the most reliable results.

The two parameters that have to be determined are: (1) Available wood stock per hectare along the river including fresh and instream wood; and (2) area adding to the driftwood supply during a flood. Comprehensive studies for a given catchment were carried out by e.g. Rimböck and Strobl (2001), who determined a specific wood stock for various driftwood input mechanisms and identified bank erosion and sliding areas using hydraulic analyses, geological surveys, and aerial views. Mazzorana et al. (2009) presented a method for generating hazard index maps along mountain torrents with the catchment classified according to its likelihood to deliver driftwood.

A detailed catchment analysis for the Sihl River was conducted by Flussbau AG (2009). They determined the specific wood stock along the river and determined the reach-averaged river width for various flood discharges using the methods of Parker (1979) and Yalin (1992). This resulted in possible rates of bank erosion, if no bank protection exists. Possible sliding areas reaching the river and their dimensions are estimated using existing hazard maps and geological maps.

2.4 *Evaluation of past flood events*

Past flood events where driftwood transport occurred, give good indications on the likelihood of a catchment to deliver driftwood. After a flood event, the 'lost' wood along a river

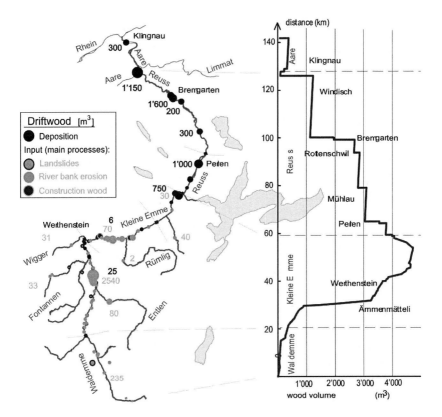

Figure 1. Driftwood balance sheet of Kleine Emme and Reuss River catchment areas determined after 2005 flood event in Switzerland and corresponding driftwood transport diagram (Source: Waldner et al. 2010, adapted).

119

may be determined using e.g. aerial views or with a site inspection. Furthermore, the volume of retained or deposited wood along the river can be estimated. After the 2005 flood event in Switzerland, Waldner et al. (2010) carried out an extensive study to document driftwood depositions and accumulations and their compositions. They reconstructed the various driftwood input and deposition processes and established driftwood balance sheets for three catchment areas. An example for the catchment of the River Kleine Emme and the River Reuss is shown in Figure 1, along with the driftwood transport diagram that is discussed in the next chapter. The driftwood balance directly indicates the hazard zones in the catchment area and may be used to select an optimum location for a future driftwood retention structure (Hunziker et al. 2009).

2.5 Driftwood transport diagram

The potential driftwood volume derived from one of the above methods is a basic parameter to design the driftwood retention structure. To choose an optimum location and to understand the effect of the retention structure, the driftwood transport diagram applies (Hunziker et al. 2009). Figure 1 shows this diagram for the Kleine Emme and Reuss Rivers. The driftwood transport diagram also indicates endangered structures, where blocking has to be expected during a flood event. The driftwood transport diagram can also be established prior to a flood event using the driftwood volume determined from the catchment area evaluation. Once the retention structures have been planned, they can be included in the transport diagram to assess their effect on the driftwood transport in the catchment area (Flussbau AG 2009).

3 DRIFTWOOD BLOCKING

Driftwood blocking at river crossing structures is the main process that triggers backwater rise and flooding of adjacent areas. A comparatively small driftwood amount or just one single log jamming due to poor structural design may recruit other logs and lead to an extensive driftwood accumulation. In the course of the hazard analysis, all river crossing structures should therefore be critically evaluated regarding their blocking sensitivity. The bridge width and height, e.g. the available freeboard for the design flood, have to be compared with the expected driftwood dimensions. In general, a representative log length and rootstock diameter is selected, depending on the existing forest in the catchment area.

Most studies dealing with the blocking probability at bridges and weirs originate in small scale model tests. One of the first systematic flume experiments to investigate the processes of drift accumulation in mountainous rivers was presented by Bezzola et al. (2002). Further results based on these model tests were presented by Schmocker and Hager (2011). Gschnitzer et al. (2013) further investigated the resulting backwater rise due to bridge clogging. The blocking at weirs and spillways was investigated by e.g. Johansson and Cederström (1995), Hartlieb (2012) or Pfister et al. (2013).

The relevant factor influencing the accumulation was often found to be the log length (Diehl 1997). Therefore, the evaluation of the catchment should always include information on the driftwood characteristics, especially log lengths, log diameters and rootstock dimensions. The blocking probability for single wood pieces generally increases with (1) decreasing Froude number; (2) decreasing freeboard; and (3) increasing driftwood dimensions. Driftwood clusters are more prone to get blocked than single pieces, and the amount of driftwood affects the blockage risk. An increased blockage risk was stated for trapezoidal cross-sections and lateral abutments. Single rootstocks are more likely to get blocked than single logs and the maximum blocking probability was observed for a drift cluster containing rootstocks. Bridge constructions with open truss structures and superstructures with open parapets, supply cables, and pipes favor the drift accumulations (Bradley et al. 2005). 'Smooth' bridge characteristics and especially a baffle construction often assure the safe passage of transported driftwood (Schmocker and Hager 2011). Countermeasures to induce a safe driftwood passage can be found in e.g. Bradley et al. (2005) or Lange and Bezzola (2006).

4   DRIFTWOOD RETENTION

Existing drift wood retention schemes are mostly designed and optimized for steep mountain torrents and do not apply for larger pre-alpine rivers. Nets or V-shaped racks span typically over the whole river width and may lead to a complete interruption of the sediment transport, leading to erosion problems and ecological deficits further downstream. In addition the mechanical load for a structure installed over the complete river width is difficult to handle in larger rivers. To avoid these problems, the so called bypass retention scheme has been developed (Schmocker and Weitbrecht 2013).

4.1   *Bypass retention*

The idea of the bypass retention scheme is to retain the driftwood in a bypass channel located along the river. It should be located at an outer river bend, where a separation of bed-load and driftwood results due to the secondary currents induced by the river curvature (Fig. 2). Driftwood tends to float towards the outer bend due to inertia and the secondary currents whereas bed-load remains at the inner bend. The outer river bank is replaced by a side weir of a smaller height than the original bank but a higher elevation than the river bed. This ensures that the bed-load remains in the river during bed forming discharges and that the bypass channel is solely active during larger flood events. The driftwood rack is located longitudinally between the main river and the bypass channel resulting in an almost parallel approach flow to the rack. In the following sections, two case studies are described. In the Kleine Emme example, the discharge towards the bypass channel is regulated with a controllable weir in contrast to the explained system above. In the Sihl River example, the discharge towards the bypass channel is regulated with a non-controllable side weir as shown in Figure 2.

4.2   *Driftwood retention at Kleine Emme*

During the 2005 flood, three bridges were destroyed and several bridges and weirs have been damaged along Kleine Emme River due to major driftwood transport. The total damage along Kleine Emme has been estimated to approx. 100 million CHF (Bezzola and Hegg 2008). The driftwood was transported further downstream and led to additional problems in Reuss River. To improve the situation, Hunziker et al. (2009) proposed the removal of at

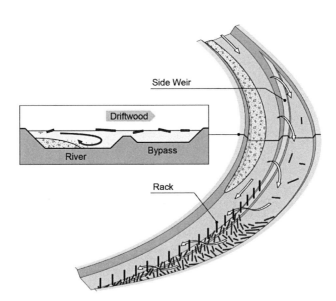

Figure 2.   Schematic view of bypass retention (Source: Schmocker and Weitbrecht 2013).

least 50% of the transported driftwood during a flood event with a bypass driftwood retention scheme. The loosely packed driftwood volume during a 100-years flood with a maximum discharge of 640 m³/s was estimated to 10,000 m³ (Hunziker et al. 2009).

Ettisbühl community was selected as the suitable location for the driftwood location, where Kleine Emme follows a distinctive left-hand bend. The location was also chosen by Steiner Energie AG for the installation of a small-scale hydropower plant. The latter was therefore combined with the bypass retention (Fig. 3a). To control the hydropower plant during regular flow conditions and the driftwood bypass during flood events, two weirs were installed, namely one in the existing river along with the hydropower station, and the other as an outlet to the driftwood bypass at the outer river bend. The driftwood weir is always closed under normal operation and opened during a flood event only, so that driftwood transported along the outer bend is guided into the bypass section. To ensure that the sediment transport still takes place along the left-hand side in the main channel, the vertical position of the bypass weir sill is 0.5 m higher than the sill of the main weir (Weitbrecht and Rüther 2009). A first design was proposed by IM Maggia AG and subsequently tested and optimized with physical scale model experiments at VAW (Tamagni et al. 2010).

In the first model tests, a V-rack located at the end of the bypass channel was tested, resulting in a backwater rise of more than 3 m, which was not acceptable in terms of flood safety. The rack was therefore placed parallel to the flow as shown in Figure 4, resulting in a strongly improved situation. The driftwood passes the driftwood weir, floats along the rack and deposits in the downstream rack corner. The current through the driftwood weir further induces a large recirculation zone, where a considerable amount of driftwood gets trapped (Fig. 3b). The driftwood accumulation density in front of the rack is therefore considerably

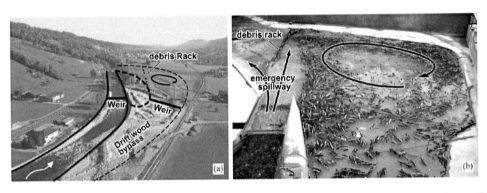

Figure 3. (a) View of left-hand river bend at Kleine Emme after 2005 flood event with scheme of planned hydro-power/driftwood retention structure, (b) HQ$_{300}$ flood experiment: Driftwood accumulation mostly at downstream corner of debris rack. A major part of driftwood is trapped in recirculating zone.

Figure 4. (a) View of bypass retention after completion, (b) Bypass section during 2013 flood event. (Sources: vif Canton Lucerne, VAW).

decreased, resulting in a backwater rise of only 1.2 m. For the worst case scenario of a completely blocked debris rack, an emergency spillway is located at the upstream end of the debris rack (Fig. 3b). As small amounts of driftwood may pass through the spillway, a second driftwood rack was placed further downstream. The final design resulted in driftwood retention of at least 60% for each tested scenario. The new driftwood potential in the Kleine Emme River and consequently in the Reuss River can now be determined by including the retention structure in the driftwood transport diagram in Figure 1. The driftwood retention structure along with the power plant was built in 2011 and covers an area of about 4 ha (Fig. 4a). The cost of the driftwood retention structure was about 7.3 Mio. CHF.

In June 2013, a small flood event occurred where the bypass channel was flooded for the first time. The transported driftwood was successfully guided into the bypass section and approximately 500 m³ were retained with an accumulation pattern similar to the scale model tests. The driftwood accumulated along the rack and especially at the downstream corner (Fig. 4b). Driftwood was also deposited in the adjacent embankments, indicating that the recirculation zone was present in the prototype as well.

4.3 *Driftwood retention at Sihl*

Two recent flood events at River Sihl in Zurich in 2005 (30-years flood event) and in 2007 (10-years flood event) highlighted the need to improve the existing flood management plans. Given that River Sihl is prone to driftwood transport, the inundation risk may significantly increase due to driftwood blockage, especially at the culverts where River Sihl passes below Zurich central train station. A detailed survey of the catchment area resulted in possible driftwood volumes of 6,000 m³ for a 100-years flood, and 12,000 m³ for a 300-years flood (Flussbau AG 2009), respectively. Therefore the bypass driftwood retention at River Sihl was developed by both the Office of Waste, Water, Energy and Air of Zurich (AWEL), the engineering company Basler & Hofmann AG, and optimized in scale model tests at VAW.

Compared to the driftwood retention at River Kleine Emme, the bypass retention at River Sihl is not regulated. The driftwood retention structure is planned in a distinctive right-hand bend, where the river width is about 40 m. As the available space is restricted by a highway, the radius of the right-hand bend is narrowed and the river stretch is moved to the right (Fig. 5). The old river channel is then used as driftwood retention area. Along the right-hand bend, a fixed weir separates the new river channel from the bypass channel. The driftwood

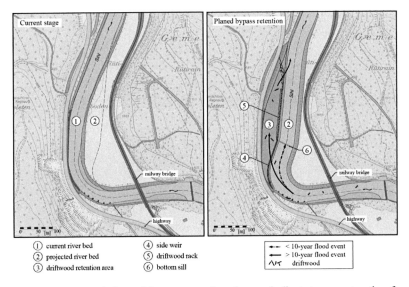

Figure 5. Current stage and planned bypass retention. Arrows indicate transport paths of water and driftwood.

123

rack is placed parallel between the old and the new river beds. A bottom sill located at the end of the bend assures that the bed level of the right-hand bend remains stable.

As no active elements are planned and intervention during a flood is impossible, the feasibility of the bypass retention depends essentially on the sound engineering design of all key elements. As no sufficient assessment was possible by means of analytical or numerical approaches, the bypass retention was investigated using a 1:40 scale model (VAW 2012).

The key elements investigated were: (1) General layout of River Sihl along bypass retention; (2) Design and especially height of side weir; and (3) Bypass section and driftwood rack. The main findings and optimizations are listed below:

– To increase the bed-load transport along the rack, the width of the bottom sill and the river width along the driftwood rack were kept constant at 30 m. The basis for this width results from the reach-averaged width (Yalin 1992) for a 30-years flood discharge.
– To successfully entrain driftwood floating along the right-hand side, the secondary currents were increased with an artificial embankment located at the inner side of the right-hand bend (Fig. 6a). The embankment decreases the cross-sectional area, resulting in increased overflow depths over the side weir especially for smaller discharges during a rise or recession of a flood. The artificial embankment reduces the initial width to depth ratio from 10–15 to 5–10. The maximum artificial embankment height is 2.0 m above the river bed. The embankment is completely submerged for discharges exceeding the 30-years flood peak.
– The side weir determines at which stage the bypass section becomes active. If the side weir is too low, bed-load may be entrained into the bypass section. If the weir is too high, especially driftwood is not entrained during the rise or recession of the flood hydrograph. To guarantee sufficient overflow depth for a 30-years flood, the side weir height was set equal to the water level of a 10-years flood, as derived from a numerical 1D HEC-RAS computation. This results in a side weir height of 2.0 m above the river bed and overflow depths for a 30-years peak discharge of approximately 0.5 to 0.8 m. For a 300-years peak discharge, the overflow depth is approximately 1.8 m at the start of the side weir and 1.3 m at its end.

Figure 6. (a) Artificial embankment located along inner river bend to increase secondary currents, (b) Flow conditions along side weir for small overflow discharge, (c) Driftwood retention after a 300-years flood event with retention efficiency of 95%.

- Logs tend to deposit on the side weir itself for small overflow depths. In the worst case, deposited logs prevent the transport of further driftwood into the bypass section. The side weir was therefore designed with a short, 0.8 m long crest, and a 0.5 m high drop to guarantee that logs tumble into the bypass section or, if they are deposited on the crest, get remobilized as the flow depth increases (Fig. 6b).
- The bypass section is about 2.0 m higher than the bed of River Sihl with a slope of 0.4%. To prevent the water from directly flowing back into River Sihl, an additional 1.0 m high embankment was placed along the upper 135 m. This guarantees sufficient flow depth for driftwood transport under small bypass discharges. The upper rack poles were placed on this embankment. The lower rack portion is 100 m long with the racks directly placed on the bypass channel bed. The retention area is approx. 6,200 m².
- The clear rack width c was estimated according to Lange and Bezzola (2006) who stated that for a given c, logs with a length of $L \geq 1.5c$ are retained. Flussbau AG (2009) stated that more than two thirds of the expected driftwood at River Sihl is 5 m or longer. The clear width of the debris rack was therefore set to $c = 3.3$ m. The poles have a diameter of 0.40 m.
- The maximum backwater rise measured in the bypass channel during the peak discharge of an extreme flood event ($Q \approx 550$ m³/s) was 4.0 m, which led to a rack height of 4.5 m. The additional freeboard of 0.5 m covers uncertainties especially regarding the porosity of the driftwood accumulation, as small branches and leaves were not modeled in small scale.

Due to this adaptions, the driftwood efficiency of the small scale bypass retention at Sihl River could be increased from originally 60% to 95% for all tested flood events (Fig. 6c). The achieved driftwood retention efficiency is therefore similar to a rack placed across the complete river cross-section. The bypass channel was by two-thirds filled for both the 300-years flood and the extreme flood, with a certain buffer provided if this volume should be exceeded. This buffer zone further guarantees that the resulting backwater rise from the drift-wood rack does not affect the flow separation at the side weir.

In general, there is a conflict of goals regarding driftwood retention and bed-load transport. The more water is diverted into the bypass channel to increase the driftwood retention, the more sediment deposits in the river. Sediment depositions were observed upstream of the bypass retention and along the debris rack. Especially the depositions upstream of the retention structure resulted in an increase of the water levels in the Sihl River as compared with the current stage. A sediment management plan is therefore required to restore the original river bed after major floods.

5 CONCLUSIONS

During recent flood events in Switzerland, driftwood transport was not limited to mountain torrents but occurred as well in larger pre-alpine rivers. Given the numerous endangered bridges and weirs, as well as the damage potential in populated areas, risk analysis and engineering measures are key elements to minimize the potential damage. One important input parameter for a hazard evaluation is the amount of transported driftwood for a given flood event. As the application of existing empirical methods for mountain torrents to pre-alpine rivers involves large uncertainties, a detailed study of the catchment area is therefore necessary for project planning. Additional data may be obtained from the evaluation of past floods.

The determined driftwood volume may be transferred into a driftwood transport diagram that indicates the entrainment locations and the potential driftwood volume along a river. It enhances the hazard evaluation and indicates preferable locations for targeted driftwood retention. An analysis of all river crossing structures regarding their driftwood blocking probability helps to identify endangered structures prior to a flood event.

The common driftwood retention structures are not suitable for pre-alpine rivers due to the resulting backwater rise and sediment retention. A new driftwood bypass retention scheme was therefore developed and optimized using hydraulic model tests. The driftwood is directed into a bypass channel located at the outer river bend and retained with a rack placed parallel to the river axis. A regulated bypass retention was built at the River Kleine Emme, indicating a good performance during a first small flood event. An unregulated bypass retention is planned at Sihl River to be built in 2015.

ACKNOWLEDGMENT

The authors would like to thank Canton of Lucerne and AWEL of Canton Zürich for both financial and technical support.

REFERENCES

Bezzola, G.R., Gantenbein, S., Hollenstein, R., Minor, H.-E. 2002. Verklausung von Brückenquerschnitten [Log jam at bridges]. Intl. Symp. Moderne Methoden und Konzepte im Wasserbau *VAW-Mitteilung* 175, 87–97, H.-E. Minor, ed., ETH Zurich, Zürich.

Bezzola, G.R., Hegg, C. (eds.) 2008. Ereignisanalyse Hochwasser 2005 Teil 2: Analyse von Prozessen, Massnahmen und Gefahrengrundlagen [Event analysis of the 2005 flood event]. Federal Office for the Environment FOEN, Swiss Federal Institute for Forest, Snow and Landscape Research WSL, *Umwelt-Wissen* 0825, WSL, Birmensdorf.

Böhl, J., Brändli, U.B. 2007. Deadwood volume assessment in the third Swiss National Forest Inventory: Methods and first results. *Eur. J. For. Res.* 126(3), 449–457.

Bradley, J.B., Richards, D.L., Bahner, C.D. 2005. Debris control structures: Evaluation and countermeasures. U.S. Dept. Transportation, Federal Highway Administration *Report* No. FHWA-IF-04–016, Washington, D.C.

Brassel, P. Brändli, U.B. (eds.) 1999. Schweizerisches Landesforstinventar. Ergebnisse der Zweitaufnahme 1993–1995 [Results from Swiss National Forest Inventory 1993–1995]. Federal Office for the Environment FOEN, Swiss Federal Institute for Forest, Snow and Landscape Research WSL, *Report*, Bern, Stuttgart, Vienna, 442 p.

Diehl, T.H. 1997. Potential drift accumulation at bridges. U.S. Dept. Transportation, Federal Highway Administration *Report* No. FHWA-RD-97-028, Washington, D.C.

Flussbau AG. 2009. Schwemmholzstudie Sihl [Driftwood potential Sihl River]. *Report* Amt für Abfall, Wasser, Energie und Luft des Kanton Zürichs, Zürich, 87 p.

Gregory, K.-J., Davis, R.J., Tooth, S. 1993. Spatial-distribution of coarse woody debris dams in the Lymington Basin, Hampshire, UK. *Geomorphology* 6(3), 207–224.

Gschnitzer, T., Gems, B., Aufleger, M., Mazzorana, B., Comiti, F. 2013. Physical model tests on bridge clogging. Proc. of 2013 *IAHR World Congress*, Chengdu, China.

Hartlieb, A. 2012. Large scale hydraulic model tests for floating debris jams at spillways, *2nd IAHR Europe Congress* "Water infinitely deformable but still limited", 27.–29. June 2012, München.

Hunziker, R., Kaspar, H., Stocker, S., Müller, D. 2009. Schwemmholz-Management Kleine Emme und Reuss [Driftwood Management at Kleine Emme and Reuss River]. *Wasser, Energie, Luft* 101(1), 21–25.

Johansson, N., Cederström, M. 1995. Floating debris and spillways. Proc. Intl. Conf. *Hydropower*, San Francisco, California, United States, 2106–2115.

Keller, E.A., Swanson, F.J. 1979. Effects of large organic material on channel form and fluvial processes. *Earth Surf. Proc. Land.* 4(4), 361–380.

Lagasse, P.F., Clopper, P.E., Zevenbergen, L.W., Spitz, W.J., Girard, L.G. (2010). Effects of debris on bridge scour. NCHRP *Report* Nr. 653. TRB, National Research Council, Washington, D.C.

Lange, D., Bezzola, G.R. 2006. Schwemmholz: Probleme und Lösungsansätze [Driftwood: Problems and solutions]. *VAW-Mitteilung* 188, H.-E. Minor, ed., ETH Zurich, Zurich.

Mazzorana, B., Zischg, A., Largiader, A., Hübl, J. 2009. Hazard index maps for woody material recruitment and transport in alpine catchments. *Nat. Hazards Earth Syst. Sci.* 9(1), 197–209.

Parker, G. (1979). Hydraulic geometry of active gravel rivers. *Journal Hydraul. Div.*, ASCE, 105(HY9), 1185–1201.

Pfister, M., Capobianco, D., Tullis, B., Schleiss, A.J. 2013. Debris-blocking sensitivity of piano key weirs under reservoir-type approach flow. *Journal Hydraul. Engng.* 139(11), 1134–1141.

Piégay, H., Gurnell, A.M. 1997. Large woody debris in river geomorphological patterns: Example from S.E. France and S. England. *Geomorphology* 19(1–2), 99–116.

Rickenmann, D. 1997. Schwemmholz und Hochwasser [Driftwood and Floods]. *Wasser, Energie, Luft* 89(5/6), 115–119.

Rimböck, A., Strobl, T. 2001. Schwemmholzpotential und Schwemmholzrückhalt am Beispiel Partnach/Ferchenbach (Oberbayern) [Driftwood potential and retention at Partnach/Ferchenbach]. *Wildbach-und Lawinenverbau* 145(65), 15–27.

Schmocker, L. Hager, W.H. 2011. Probability of drift blockage at bridge decks. *Journal Hydraul. Engng.* 137(4), 480–492.

Schmocker, L., Weitbrecht V. 2013. Driftwood: Risk analysis and engineering measures. *Journal Hydraul. Engng.* 139(7), 683–695.

Tamagni, S., Weitbrecht, V., Müller, U., Hunziker, R., Wyss, H., Kolb, R., Baumann, W. 2010. Schwemmholzrückhalt Ettisbühl/Malters [Driftwood retention Ettisbühl/Malters]. *Wasser, Energie, Luft* 102(4), 169–274.

Uchiogi, T., Shima, J., Tajima, H., Ishikawa, Y. 1996. Design methods for wood-debris entrapment. *Intl. Symp. Interpraevent* 5, 279–288.

VAW 2008. Ereignisanalyse Hochwasser 2005: Teilprojekt Schwemmholz [Analysis of 2005 flood event: Sub-project driftwood]. *Report* 4240. Laboratory of Hydraulics, Hydrology and Glaciology, ETH Zurich, Zürich.

VAW 2012. Schwemmholzrückhalt Sihl—Standort Rütiboden [Driftwood retention Sihl River—Location Rütiboden]. *Report* 4293. Laboratory of Hydraulics, Hydrology and Glaciology, ETH Zurich, Zürich.

Waldner, P., Köchli, D., Usbeck, T., Schmocker, L., Sutter, F., Rickli, C., Rickenmann, D., Lange, D., Hilker, N., Wirsch, A., Siegrist, R., Hug, C., Kaennel, M. 2010. Schwemmholz des Hochwassers 2005. *Final Report*. Federal Office for the Environment FOEN, Swiss Federal Institute for Forest, Snow and Landscape Research WSL, Birmensdorf.

Weitbrecht, V., Rüther, N. 2009. Laboratory and numerical model study on sediment transfer processes in an expanding river reach. *33rd IAHR Congress*, Vancouver (CD-Rom).

WSL 2006. Einfluss ufernaher Bestockungen auf das Schwemmholzaufkommen in Wildbächen [Effect of bankside wood on the driftwood potential]. *Report*. Federal Office for the Environment FOEN, 95 p.

Yalin, M.S. 1992. *River mechanics*. Pergamon Press, Oxford and New York.

Swiss Competences in River Engineering and Restoration – Schleiss, Speerli & Pfammatter (Eds)
© 2014 Taylor & Francis Group, London, ISBN 978-1-138-02676-6

Design of a bed load and driftwood filtering dam, analysis of the phenomena and hydraulic design

M. Bianco-Riccioz & P. Bianco
IDEALP SA, Sion, Switzerland

G. De Cesare
EPFL-LCH, Lausanne, Switzerland

ABSTRACT: Flood protection often calls on to the realization of retention works for bed load as well as wood and debris flow. Certain relatively recent arrangements did not perform according to their intended function, what shows the complexity of the design and the implementation of such works. Adaptations were necessary to reach the security objectives.

The design of a retention dam for solid materials and floating driftwood requires the consideration of numerous hydraulic and material transport processes. The analyses and design validation can be made with two approaches: physical modelling by the construction of a reduced scale model and the test realization or numerical simulation, by means of software packages such as GESMAT (1D) or TOPOFLOW (2D). The present work consists in implementing both approaches, in estimating and in comparing the answers which could be given for a bed load and debris flow filtering dam on a river with a slope of the order of 10%.

Thanks to water level gauges and visual observations during tests on the physical model, the progression of the obstructions by driftwood and bed load is well understood, and the effectiveness of these obstructions proven. The tested work plays at first a role of filtering and retention and secondly a role of side overflow towards a zone with low damage potential, when the capacity of the in-stream retention space is reached.

The performed numerical simulations, essentially in 1D, reproduce well the phenomena of bed load aggradation. Moreover, the potential obstruction by floating wood is considered and influences the behavior of the structure.

By putting in parallel physical and numerical models, it was possible thanks to the results from the physical scale model to refine the numerical simulation tools taking into consideration additional components and behavior-type rules. These further established rules can now be used for other cases where physical modelling is not foreseen.

1 INTRODUCTION

The issue of flood retention, the control of floating driftwood and bed load transport in an alpine environment is an important topic in the field of torrential hydraulics. Flood hazard maps exist and indicate the habitations threatened by flooding. In these threatened zones, there is a corresponding risk, more or less important, of loss of human lives. There is thus an absolute necessity to reduce the danger to a level judged acceptable.

In an alpine environment, the phenomena of the bed load transport play an important, even dominant, role in the determination and the establishment of flood hazard maps. Indeed, houses and other infrastructures in connection with human activities are often located on alluvial cones where the slope diminishes and where bed load transported by the torrent has thus the tendency to accumulate. This accumulation of solid materials leads to a decrease of the hydraulic capacity of the channel which results in a greater flood risk.

The necessity of designing structures which allow for the retention and/or the filtering of transported sediment and driftwood is evident. These works allow for avoiding a decrease or loss of the hydraulic capacity of the channel caused by excessive sediment deposits, by obstruction of bridges or by clogging of the channel due to driftwood.

Tests on a physical model together with numerical simulations of a real case, i.e. an alpine torrent in central Valais, Switzerland, with a slope of approximately 10% allowed (a) the evaluation of its functioning, (b) the proposition of alternative structural dimensions as well as (c) recommendations for the implementation and the optimization of this type of hydraulic work.

The tests on a physical model allowed for numerous analyses of the behaviour and the possibilities to build such a structure. The effect of driftwood could be well highlighted and rules were established and used for the numerical modelling.

The numerical simulations permitted to reproduce the functioning of the structure and allowed for adjustments. The solid and liquid flow behaviour were identified and quantified. The numerical simulation take into account the phenomena of obstructions by driftwood, their occurrence and time evolution as well as the conditions in which they occur. The GESMAT simulation program could in this way be extended by new hydraulic components.

2   OBJECTIVES

A retention structure for solid and liquid flows was planned to improve the flood safety on an alluvial fan, where the torrent slope varies between 6 and 9%. Based on the project of a filtering dam, the present work consisted of carrying out the physical model tests, as well as one-dimensional and two-dimensional numerical simulations using IDEALP in-house software packages.

The objectives are the evaluation of the functioning of a "slit filter dam", the proposition of different geometries and dimensions as well as recommendations for the realisation and the optimisation of this type of structure. The physical and numerical models should be able to highlight the uncertainties as well as the adequacy of the different methods used in the predesign phase.

The main objectives of the study are the comprehension of the encountered phenomena as well as the hydraulic dimensioning of the various elements of the filtering dam, based on the analysis of the numerical and physical model test results. This study allows engineers to recognize the value of the different analysis and calculation tools.

There are many elements of the filtering dam to be conceived, designed and tested regarding their functioning:

– Bottom outlet which allows regular floods to pass
– Vertically arranged openings generally equipped with lateral beams made out of steal
– Adequate spacing of the lateral beams of the openings
– Overflow spillway on the top of the dam
– Stilling basin at the downstream end of the spillway
– Side spillway to evacuate the case of flood overload occurrence and to divert the discharge towards an additional retention space with a low damage potential.

The main questions that arise are:

– The time of obstruction of the bottom outlet
– The lateral overflow of water, debris and bed load in case of
   1. The influx retention volume is full and/or
   2. The trash rack is obstructed with driftwood and/or
   3. The overflow spillway capacity is exceeded
– The quantities of water, debris and bed load that still transit downstream.

3 PHYSICAL MODELLING

3.1 *Type and scale of the model*

The extent of the study zone covers an area of 150 m by 350 m at prototype scale, which corresponds at the physical model scale to approximately 5 m by 12 m, built directly at natural ground level with corresponding slope next to a vineyard.

The physical model is built at the geometrical scale of 1:30. It is operated respecting Froude similarity, by admitting the preservation of the ratio between inertia and gravity forces. Figures 1 and 2 give an idea of the configuration of the physical model structure and its various components.

3.2 *Components of the model*

3.2.1 *Floating debris screen*

A floating debris screen situated upstream of the overflow dam allows for the retention of driftwood and favor sediment deposit in the influx retention space.

Figure 1.   Physical model of the filtering dam with retention zone, view from upstream.

Figure 2.   Physical model of the filtering dam with stilling basin, view from downstream.

One of the objectives is to determine the best configuration that retains the driftwood without totally blocking the trash rack and not leading rapidly to the case of lateral flood overload. Lateral flood overload should not occur before reaching a flood event with a 100-year return period.

3.2.2 Orifice

The narrowing of the flow section, with the bottom outlet and the transversal bars across the opening, provokes a backwater effect that is necessary to retain sediments, in particular in case of floods with few floating wood.

The bottom opening and the spaces between the lateral beams can be blocked by the floating debris that manage to cross the trash rack.

3.2.3 Overflow spillway on the top of the dam

When the water level reaches the level of the dam, the overflow spillway is activated.

Generally, the overflow spillway as well as the remaining opening has to allow for the transition of the entire discharge downstream without the risk of erosion nor the risk of dam failure.

Figure 3. Floating debris screen installed in the physical model.

Figure 4. Bottom opening and upper flow section with horizontal beams on the physical model.

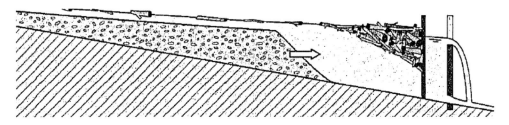

Figure 5. Overflow and gradual deposition of bed load and driftwood retention during an important flood event.

Figure 6. Stilling basin and downstream sill of the physical model.

If there is a possibility of lateral overflow, as it is the case in the studied configuration, this side overflow prevents the uncontrolled flooding of the dam and plays the role of a safety work reducing further the risk of erosion or dam failure.

3.2.4 *Stilling basin*
Downstream of the spillway, the stilling basin dissipates the flow energy. It is made of reinforced rockfill at its upper limit, then of a double layer of large-dimension freestanding blocks. This stilling basin is limited downstream by a sill at the level of the existing river bed. It ensures the safe transition of the flood discharge through the orifice and over the spillway, the dissipation of the energy and prevents erosions of the dam toe.

The geometry of the stilling basin and the height of the downstream sill permit additional filtering of bed load materials. In case of excess bed load transport, the sediment deposits create a reduced longitudinal slope that blocks the bottom opening and improves the retention conditions upstream of the dam.

3.2.5 *Lateral overflow*
For rare floods up to extreme events, according to the planned configuration, after the retention basin formed by the natural river bed has filled up with sediment deposits to the same

133

elevation of the right bank, side overflows will occur. The flow will then orientate towards a zone with a minor damage potential, which was converted into a zone for additional sediment deposits for this purpose.

3.3 Simulations carried out

Floods of various return periods were simulated in the physical model, i.e. 10-year, 30-year, 100-year return periods and extreme flood events. All tests were performed by adding driftwood as well as solid materials, except for the 10-year-flood scenario.

4 NUMERICAL SIMULATIONS

4.1 GESMAT model

The GESMAT model is a one-dimensional (1D) hydraulic program modelling bed load transport by fractions and which has a quasi-stationary module (sequences of backwater curves) as well as a non-stationary module by solving the Saint-Venant shallow water equations. The program is designed for small slopes (inferior to one per mill) as well as for steep slopes (up to 20%). It allows for the consideration of numerous hypotheses concerning the development of an armoring layer as well as the presence of different artificial structures.

The components of the filtering dam are implemented in terms of the outlet (i.e. orifices or culverts), lateral beams and overflow structures. The partial or total obstruction of the orifice can be controlled by defining the speed of obstruction. The program enabled the numerical simulations of scenarios that are identical to the ones used in the physical model.

4.2 TOPOFLOW model

The model TOPOFLOW is a two-dimensional (2D) hydraulic model with implemented modules for the calculation of the bed load transport. It also allows investigating a wide range of slopes and influences from artificial structures. Additionally, the application of a 2D model enables the observation of the evolution of the riverbed morphology as well as lateral overflow. Good agreement between the geometry of the calculated and measured deposits has been observed. However, the present article is limited to the detailed comparison of the one-dimensional numerical model GESMAT and physical model.

5 COMPARISON OF THE RESULTS

5.1 The 10-year-flood scenario (Q10)

The 10-year scenario was carried out without adding driftwood. The discharges correspond to a constant flow of 8 m³/s over 55 minutes. The deposits in the upstream zone of the trash rack calculated by GESMAT (1.10 m) and the ones measured after the physical test (1.20 m) are quite similar. The bed load transport was reduced by the filter dam in the order of about 30% in both cases, the numerical simulation as well as the physical model.

5.2 The 30-year-flood scenario (Q30)

The 30-year scenario was carried out with adding driftwood. The discharges correspond to a constant flow of 16 m³/s over 55 minutes. The deposition height calculated by GESMAT (3.60 m) and the one measured after the physical test (3.90 m) are again comparable.

The bed load transport was reduced by the filter dam in the order of about 80% in both cases. The numerical model does not indicate sediment depositions at the bottom opening.

Figure 7. Situation at the dam after 10-year-flood event (physical model, flow from left to right).

Figure 8. Situation at the dam after 30-year-flood event (physical model, flow from left to right).

This result was confirmed by the observations on the physical model where also no obstruction of the orifice had been observed.

5.3 *The 100-year-flood scenario (Q100)*

For the prototype, the peak discharges of a 100-year flood are about 30 m³/s. The rising time is about 3 hours and the decreasing part takes about 6 hours. As indicated in Figure 9, the evolution of the riverbed calculated by GESMAT and measured after the test in the physical model are again alike.

The bottom outlet is blocked at the end of the test (Fig. 11). For the numerical model, the obstruction was defined by the opening between the lowest bar and the riverbed being inferior to 50 cm. This corresponds to the diameter of the considerably largest driftwood trunks. Figure 10 shows the differences between the water levels of the numerical and the physical model. The agreement is good, even if the dynamics of the obstruction are not the same (scale effects). Indeed, at the end of the tests, the amount of stored sediments in front of the dam is the same.

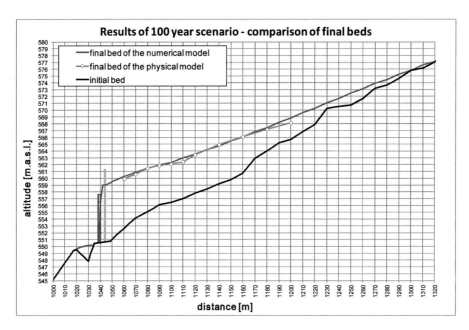

Figure 9. The 100-year-flood scenario—comparison of the sediment accumulation heights of the numerical and physical model along the thalweg.

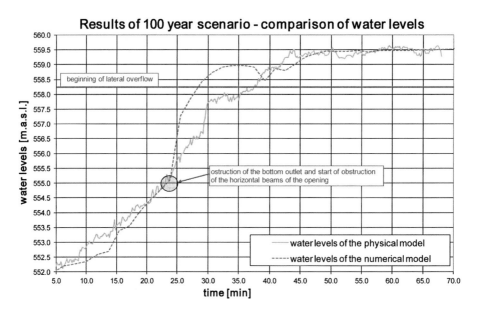

Figure 10. 100-year return period flood event—comparison of water levels for the physical and numerical model in function of time.

6 SYNTHESIS

The physical model reproduces reliably the bed load processes at slit filter dams, implicitly the clogging caused by driftwood. However, it cannot be used for each project. The numerical model reproduces properly the dynamic processes related to sediment transport, i.e. depositions in front of the dam and bed load transition towards the dam. Furthermore, the lateral

136

Figure 11.   Situation at the dam after 100-year return period flood event on the physical model.

overflow of the dam can be evaluated pertinently. Thus, the numerical model allows for the investigation of different hydrological and obstruction scenarios in order to take into account variable site conditions.There is a good agreement between the results of the numerical and physical model for all tested scenarios. In conclusion, the bed load transport and driftwood accumulations at slit filter dams can be reproduced effectively using numerical models, such as GESMAT and TOPOFLOW.

ACKNOWLEDGEMENT

The project « Conception d'un barrage filtrant—Analyse des phénomènes et dimensionnement hydraulique » has been performed partially as a HES Bachelor work, it has been financially supported by CimArk Valais.

REFERENCES

De Cesare, G. (2010). *Support de cours d'Hydraulique—hydrostatique, hydrodynamique, écoulements en charge, écoulements en nappe libre*. Yverdon: HEIG-VD.

Fehr, R. (1987). *Einfache Bestimmung der Korngrössenverteilung von Geschiebematerial mit Hilfe der Linienzahlanalyse*. Schweizer Ingenieur und Architekt 105(38): 1104–1109.

Graf, W.H. and Altinakar, M.S. (2011). *Hydraulique fluviale*. Traité Génie Civil. Vol. 16. Lausanne: Presses Polytechniques et universitaires romandes.

Jäggi, M. (1995). *Flussbau*, Vorlesungsskript, ETH Zürich, Assistenz für Wasserbau.

Jäggi, M. (1992). *Sedimentaushalt und Stabilität von Flussbauten*. VAW Mitteilung 119, ETH Zürich.

Lencastre, A. (1995). *Manuel d'hydraulique générale*. Eyrolles, Paris, 633 pages.

Nussle, D. (2009), *Bois flottants, processus, mesures d'aménagement hydraulique*, Cours de formation continue "Protection contre les crues", 19/20 mars 2009 à Granges-Paccot, Association suisse pour l'aménagement des eaux et KOHS.

Rickenmann, D. (1997). *Schwemmholz und Hochwasser*. Wasser Energie Luft 89(5/6): 115–119.

Rickli, C. and Hess, (2009). *Aspects de la formation des bois flottants*, Cours de formation continue "Protection contre les crues", 19/20 mars 2009 à Granges-Paccot, Association suisse pour l'aménagement des eaux et KOHS.

Schleiss, A.J. (2008). *Cours Aménagements Hydrauliques*, Ecole polytechnique fédérale de Lausanne.

Sinniger R.O. and Hager W.H. (2008). *Constructions Hydrauliques*. Traité Génie Civil. Vol. 15. Lausanne: Presses Polytechniques et universitaires romandes.

Swiss Competences in River Engineering and Restoration – Schleiss, Speerli & Pfammatter (Eds)
© 2014 Taylor & Francis Group, London, ISBN 978-1-138-02676-6

Design of a diversion structure for the management of residual risks using physical model tests

A. Magnollay
BG Consulting Engineers SA, Lausanne, Switzerland

G. De Cesare & R. Sprenger
EPFL-LCH, Lausanne, Switzerland

D. Siffert
Municipality of Delémont, Switzerland

P. Natale
P. Natale SA, Delémont, Switzerland

ABSTRACT: As revitalization and protection against floods are essential in our time, the town of Delémont in Switzerland has set up a major project named "Delémont Marée Basse". Part of the project, the residual risk plans to be managed by a lateral overflow bypassing the town via the railway platform. A physical model of the area is built at the scale of 1:32 in order to test and optimize the control structure. The main objective of the latter is to protect the town against extreme flood discharges. The main objectives are to derive 15 m³/s during a 300-year flood (150 m³/s) and avoid overflow for the 100-year flood (110 m³/s). After a major calibration phase of water lines, different geometries of the control structure are tested. The retained configuration is optimized for giving results consistent with the requirements of the project. The model was then used to define the minimum overflow area required, the influence of the support piles of the control structure on water lines and diverted discharge, as well as the influence of an upstream apron on Patouillet Bridge. In addition, qualitative tests examined bed load transport impact and driftwood behavior. Similarly, velocity measurements integrate the risk of erosion around the control structure and scour around the bridge piers. Finally, in preparation for the planning of future interventions, overflow discharge values and water lines are recorded under various scenarios.

1 INTRODUCTION

The Sorne River flows across the town of Delémont in a strongly channeled bed. The project "Delémont Marée Basse" (Delémont low tide) plans to manage the residual risk by a controlled overflow on the right bank downstream of the stadium named "La Blancherie". The overflow reaches the Sorne River again downstream of the town via the railway platform. The overflow structure must protect the city against flooding for high discharges. This structure, which is an important point in the residual risk management, has been tested and optimized to ensure accurate operation.

1.1 The project site

The project site is situated on the Sorne River in Delémont, just upstream of the the town center. It is situated on the last sector without any major construction and very close to the railway area. The objective of the diversion project is to use the railway installation as a discharge channel to the Sorne River downstream of Delémont.

Figure 1 presented the Sorne River in the center of Delémont. At the current stage the river is a concrete channel. The project "Delémont Marée Basse" will transform the river bed. It will allow a higher capacity and a better environmental value. Figure 2 shows the project site with the diversion on the right side.

1.2 *Hydrologic considerations*

The Sorne River is one of the tributaries of the Birse River, which joins the Rhine River in Birsfelden close to Basel. It has a catchment area of 241 km² at the Delémont gauging station. The discharges to be studied are between 90 and 165 m³/s (Fig. 3), which are respectively the values of 30-year return period flood and an extreme flood event (PMF).

1.3 *Past flood events*

Two major observed floods events have been observed in the past century, in 1973 and in 2007 (Fig. 4). In both case the observed discharge was approximately 90 m³/s.

These two events induced major flooding in the town of Delémont, although the discharge was approximately a 30-year return period flood. A free board of 0.6 m under the Patouillet Bridge during the 2007 flood has been observed, this value is useful for physical model calibration.

Figure 1. Picture of the channelized Sorne riverbed across the city of Delémont.

Figure 2. Picture of the project site with the overflow area on its right bank and a 3D view from the river.

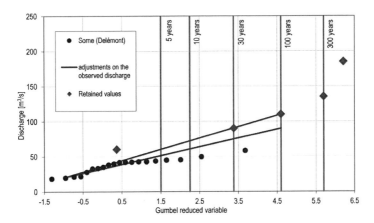

Figure 3. Flood discharge statistics according to a Gumbel distribution and relevant discharge values for corresponding return periods.

Figure 4. During the 2007 flooding in the Delémont region (Q ≈ 90 m³/s)—similar view as Figure 2.

2 PROTECTION OBJECTIVES

Currently, the capacity of the Sorne River in the center of Delémont is between 70 and 90 m³/s, but the planned developments will increase the capacity to 135 m³/s (200-year flood). The main objectives are to divert 15 m³/s during a 300-year flood (150 m³/s) and avoid over-flowing for the 100-year flood (110 m³/s).

The diversion should prevent that the discharge in the center of Delémont exceeds the 200-year flood. This value is the design flood discharge of the future Sorne training develop-ment in Delémont.

As the entire project has a revitalization purpose, the derivation scheme should have an impact as low as possible on the river bed and be located on one side of the river only. The operation must be robust and work without any human intervention or mobile parts.

3 PHYSICAL MODEL TESTS

A physical model (Fig. 5) is built at the scale of 1:32 respecting Froude similarity. It integrates sufficient upstream and downstream stretches of the Sorne River to satisfy in- and outflow conditions. The model is built with a fixed bed, levees and overflow zone.

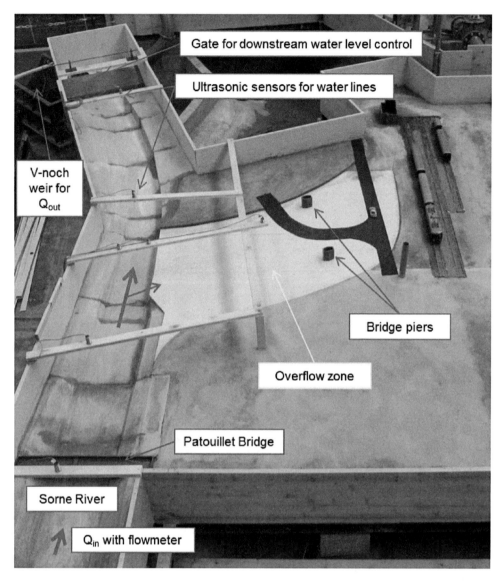

Figure 5.   Picture of the physical model in the LCH Laboratory of the EPFL and its main elements.

The lateral overflow zone and constructions influencing the flow such as pressure flow under the Patouillet Bridge, flow separation at bridge piers are also reproduced in the model. The area at the entrance of the Delémont main train station is modelled as well, where the excess flow discharge will enter the railway platform to reach the Sorne River again downstream of the town.

Ultrasonic sensors are used to measure the water lines along the Sorne River and over the overflow zone. Discharge measurements (electromagnetic flow meter and V-notch weir) are used to quantify their distribution.

After a major calibration phase of the model roughness (Baiamonte et Ferro 1997) to obtain compliance with the numerical model HEC-RAS (USACE 2013) and in situ observations during past flood events, flow lines and overflowing discharge values are listed with the Sorne at the current and project state without control structure but with the overflow zone excavated. They help identify overflows to provide the necessary level of raising the banks. This initial configuration does not satisfy the main objectives of flood diversion.

Then, two different geometries of the control structure are tested. On the second one, a number of parameters are optimized, such as the length and height of the structure. Additional qualitative tests are made on it.

3.1 Model parameters

3.2 Alternatives tested

Two main geometries of the control structure were tested. The first one (Fig. 6) is a half-moon structure, conical on the bottom at the right bank just downstream the overflow zone. The left bank creates a flow section reduction as well.

The second control structure tested (Fig. 7) is in the bed of the river, longitudinally located around the middle of the overflow zone. A smooth ramp of the right bank upstream the control structure is also realized.

3.3 Test Results and retained solutions

The first geometry of the control structure doesn't give a solution in accordance with the objectives. Its effect on the flow is strictly local and does not increase the diverted discharge rate.

The second geometry of the control structure gives results consistent with the requirements of the project. To achieve these, a number of parameters are optimized, such as the

Figure 6.   First geometry of the control structure—upstream view.

Figure 7.   Second and retained geometry of the control structure—upstream view.

length, the elevation of the lower limit and height of the structure. After optimization, only small waves pass over the overflow zone in the 100-year flood (110 m³/s). For a Sorne River discharge of 150 m³/s, the value of the diverted discharge is 15 m³/s (Fig. 8).

The model was then used to define the minimum overflow zone required, regarding the flow velocities concentration on the left side. The required area is 2 to 3 times smaller than the one initially planned, giving flexibility to adapt the excavated area to the land use plan.

A test with two support piles of the control structure in the bed (Fig. 9) shows that the derived discharge is weakly increased for discharges below 150 m³/s and that the influence of the piles for discharges above 150 m³/s is negligible.

The influence of an upstream apron on the Patouillet Bridge (Fig. 10) is positive (e.g. Jaeggi 2007) for the values of the diverted discharge, decreasing it for a discharge of 110 m³/s and increasing it for discharges over 135 m³/s.

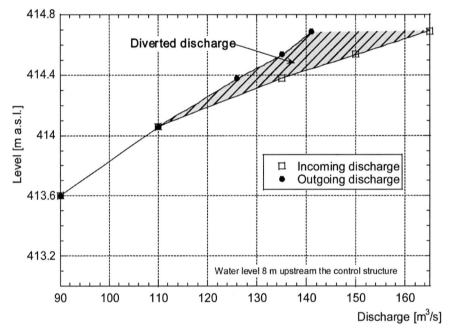

Figure 8.   Incoming, outgoing and diverted discharges measurements at the project state with control structure (geometry retained) in relation to water levels 8 m upstream the control structure.

Figure 9.   Support piles of the control structure in the river bed.

Figure 10.   Upstream apron on the Patouillet Bridge.

Figure 11.   Flow around and below the diversion structure in the physical model (Q_{in} = 135 m³/s).

Figure 12.   Driftwood around the control structure—downstream view.

At the end, additional qualitative tests showed that bed load transport affects very little the water lines, only locally by amplitude fluctuations of surface waves.

The measurements of flow velocities showed a concentration of flow on the left bank of the overflow zone, with a maximum flow velocity close to 2 m/s to a depth of 0.6 m at a flow rate of 15 m³/s. Scour around the bridge piers is a critical point and has to be taken into account during design. Similarly, these measures indicate the risk of bottom erosion downstream of the control structure and on the opposite bank. Flow around and below the diversion structure shows these high local accelerations and rather calm lateral overflow (Fig. 11).

145

The additional tests showed also, that driftwood and the sediment transport should not disturb the performances of the diversion scheme (Bezzola and Lange 2006). No blockage of the diversion structure could be observed.

Finally, in preparation for the planning of future interventions, overflowing discharge values and water lines are recorded under two different scenarios. In case the control structure is built before the implementation of measures to increase the capacity of the Sorne River through the town, the diverted discharge will increase. In case of removal of the rather heavy vegetation of the upstream riverbanks, the diverted discharge will decrease due to generally lower water levels.

4 SUMMARY AND OUTLOOK

The chosen configuration works properly. Indeed, the objective of diverting 15 m³/s during a 200-year flood (150 m³/s) is achieved. This solution seems to be robust, according to the minor influences of the geometry changes or of the bed load transport.

As a purely numerical approach would not properly represent the diversion work process with sufficient precision the physical model approach has been chosen. The possibility to show the future state of the river, its diversion structure and the simulation of a major flood event to the main stakeholders of the project (administration, politics, engineers, general public, etc.) is another great advantage.

The detailed design of the entire structure is terminated; the costs are estimated at about CHF 700'000. The project should be implemented in 2015. A landscape integration analysis of the control structure and the overflow zone has allowed suggesting some adaptation of the overflow zone that will be used as a parking lot. The slopes marking the topography of the overflow weir will be covered with lose rockfill, symbolizing the passage of water. The upstream petanque field will be renovated and open terraces facing the river will be added. The face of the control structure will be made with an engraved Corten steel plate.

REFERENCES

Baiamonte, G., Ferro, V. (1997). The influence of roughness geometry and Shields parameter on flow resistance in gravel-bed channels. Earth Surface Processes and Landforms, Vol. 22, 759–772.

Bezzola, G. R., et Lange, D. (2006), Schwemmholz—Probleme und Lösungsansätze, Mitteilungen der Versurchsanstalt für Wasserbau, Hydrologie un Glaziologie, ETHZ, Zürich.

Jaeggi, M. (2007), Sediment transport capacity of pressure flow at bridges, Proceeding of the 32nd congress of the International Association for Hydraulic Research and Engineering, July 1–6, Venice, Italy.

USACE (2013). HEC-RAS Hydrologic Engineering Centers River Analysis System, reference and users manuals.

Swiss Competences in River Engineering and Restoration – Schleiss, Speerli & Pfammatter (Eds)
© 2014 Taylor & Francis Group, London, ISBN 978-1-138-02676-6

From vision to realisation—opportunities and challenges in restoring the river Bünz

K. Steffen
Division of Hydraulic Engineering, Department of Construction, Transport and Environment of the Canton of Aargau, Aarau, Switzerland

ABSTRACT: The river Bünz is situated in the Swiss Plateau. Until the 1920s, it was a meandering river which flooded its banks regularly and its swampy flood plains could not be used for agriculture. Therefore it was decided to canalise the Bünz in a trapezoidal bed. Thus, agricultural land was gained. However, the changes produced a monotonous riverbed with barely any habitats for plants and aquatic organisms. Triggered by the floods in the 1990s and the increasing societal interest in ecological values, it was decided to restore the Bünz. The main vision was to return it to a near natural state on half of its length until 2015. Today, we are on track to achieve this target. The paper reviews and visualises the Bünz project from its initial vision to most recent work. Key elements for success or failure, a) political acceptance b) land acquisition and c) financing, are presented.

1 INTRODUCTION

As most rivers of the Swiss Plateau the river Bünz has been changed fundamentally by humans over the last 100 years. During the Bünz correction from the 1920's to the 1940's the meandering river was straightened and canalised in a trapezoidal concrete bed (Fig. 1). The floodplain was drained and many lateral tributaries were channelised and buried (UAG 2005). Thus, agricultural land was gained and flood protection was improved, facilitating the development of settlement along the waterway.

However the correction of the river Bünz resulted in a massive decline in aquatic organisms and a considerable loss of ecological valuable habitats. Due to the discharge of industrial and municipal wastewater into the river, enhanced fish mortality occurred regularly. Thus, the fish population declined significantly (UAG 2005). Therefore initial demands for a restoration came from the local fishery. The ecological deficits were not the only

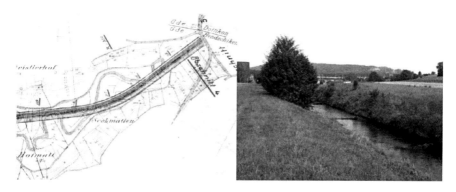

Figure 1. Plan of the correction of the river Bünz (Herzog 1918) and canalised bed after the correction (Canton Aargau 2014) in the community of Hendschiken.

problems: Urbanised areas were regularly damaged by floods. Floods, the population's increasing demand for ecological value and changes in legal requirements boosted the restoration. The Bünz is currently being reverted from a canal to a seminatural river.

This paper aims to provide an overview of the history of river restorations along the Bünz, based on experiences of involved cantonal project managers.

1.1 *Catchment area and hydrology of the Bünz*

The river Bünz is situated in the Canton of Aargau and drains a catchment area of 125 km². After 25 km it flows into the Aabach and after another 230 m into the Aare, one of Switzerland's largest rivers. The average annual discharge at the Wohlen gauge (catchment area 53.08 km²) is approximately 1 m³/s (Colenco 2002). The discharge of the 100-year flood represents 45 m³/s in Wohlen and reaches 80 m³/s before joining the Aabach (HZP 2009). The catchment area is characterised by a 2 km to 3 km wide intensive agricultural plain surrounded by two wooded ridges.

2 VISION OF A RESTORED BÜNZ

2.1 *Background*

Changing social and environmental demands in the 1990s triggered a paradigm shift towards more natural watercourses. This shift resulted in new governmental laws regarding the management of watercourses, namely the Federal Act on the Protection of Waters (Waters Protection Act, WPA) in 1991 (SAEFL/FOWG 2003). Article 37 para. 2 of the WPA requires that *the natural course of the body of water must wherever possible be preserved or restored*[1]. The legal basis for river restoration was established. When, in the beginning of the 1990s, the local fishery asked for rehabilitation of the Bünz, a first pilot study was launched. Boxes, mimicking a fish habitat, were placed in the upstream of the community of Wohlen. However, these measures contributed only to a short-term increase in fish population. Subsequently the concrete river bed in Wohlen was removed. To create a basis for future flood protection and restoration projects on the Bünz, the Canton of Aargau initiated an integrated river management study in 1993/1994. The study revealed that 88% of the morphology of the river Bünz was severely impacted. It was realised that there existed a huge need for action (Colenco & creato 1994).

2.2 *Vision*

To turn the Bünz into its original state was simply impossible, as the surroundings have changed dramatically in the preceding 100 years. Therefore the decision was to restore the Bünz as close to nature as possible, the vision being for at least half of its length to be restored to a little impacted to seminatural ecomorphological level[2]. A precondition for all restoration projects should be adequate flood protection (Colenco & creato 1994, UAG 2005).

3 OVERVIEW OF RESTORATION PROJECTS AT THE BÜNZ

Today, 11.5 km (46%) of the Bünz are categorised as being in a little impacted to a seminatural ecomorphological state. 1 km is still under construction and 800 m will be realised between

---

[1]Federal Act on the Protection of Waters (Waters Protection Act, WPA) of 24 January 1991 (Status as of 1 January 2014).
[2]Level based on the method Ecomorphology Level I (regional survey) which takes into account variability of water width, modifications of channel bed and river banks, width and structure of the riparian zone (additional information at www.modul-stufen-konzept.ch).

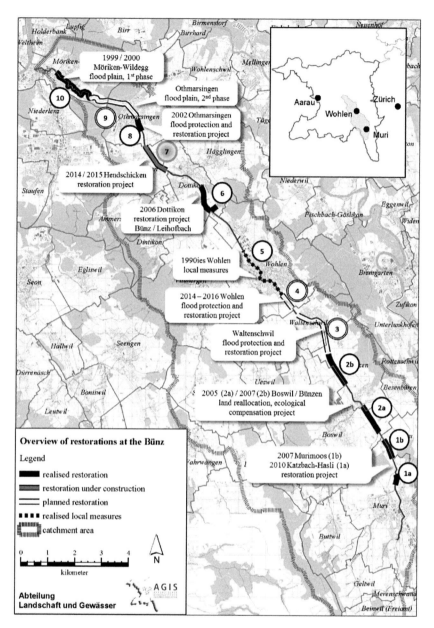

Figure 2. Overview of restoration projects along the river Bünz.

2014–2016. Another 2.9 km are planned. An overview of already implemented and yet to be planned restoration projects is shown in Figure 2 as well as in Tables 1–3.

4 FROM VISION TO REALISATION

River restoration in densely used landscape is a big challenge. The Canton of Aargau as well as several communities must give all legal authority for realisation and financing. The strategy for implementing the restorations and the key elements for success or failure such as political acceptance, land acquisition and financing were discussed with the project managers of the Bünz. The results of the discussions are presented in detail here.

Table 1.   Measures within the existing river plot: Implementation of a natural channel bed and structuration of the riverbanks.

Project Nr.	Measures	Land acquisition and space provided for waters	Trigger
1a, 1b, 5, 8	Removal of concrete bed and riprap, implementation of a natural channel bed with structures (groynes, rootstocks), reestablishment of connectivity, where possible variations of the riverbanks and local widening	No new space	1a, 1b: Proposition of WWF River Watch Programme 5: Fishing association 8: Flood protection

Table 2.   Widening and variation of the section.

Project Nr.	Measures	Land acquisition and space provided for waters	Trigger
2, 3, 4, 6, 7	Removal of concrete bed and riprap, implementation of a slightly meandering profile, widening, variations of riverbanks (shallow and steep banks) with natural riparian vegetation, structuring of bed and banks (low flow channel, rocks, groynes, rootstocks), reestablishment of connectivity	Enlargement of the existing space atleast up to recommended minimum width of 25 m, if possible to the bio-diversity width* of 37 m by land real locations and real replacement	2: Ecological compensatory measures 3: Flood protection 4: Ecological compensatory measures 6: Compensation for the preparation of forest into land for building 7: Restoration

*The minimum area is designed to ensure natural watercourse functions to the minimum extent necessary. The biodiversity area is designed to promote the natural diversity of animal and plant. (SAEFL/FOWG 2003), see also chapter 4.4.1 of this paper.

Table 3.   Reactivation of the floodplain.

Project Nr.	Measures	Land acquisition and space provided for waters	Trigger
9, 10	9: Preservation of the floodplain by acquisition of the land affected by erosion processes, removal of lines and cables out of the erosion zone 10: Extension of the floodplain, removal of the concrete bed, initiating of meandering, acquisition of the land	Enlargement of the existing area by as much as 100 m, to 400 m by land real location and real replacement	9: Creation of the floodplain during the flood of 1999 10: Floodplain protection program of Canton of Aargau

4.1   Strategic decision making: Overall project or individual project?

The study of integrated river management of the Bünz made the approval of an overall project appropriate. However reaching a political consensus with several communities in order to realise an overall project is quite a challenge. The political and financial barriers are high. For that reason, the cantonal agency decided to realise the vision stepwise through individual projects. This strategy of parallel planning seemed to be more successful because it allowed to wait for the right moment in each community. Because of flexibility, the project benefited from local opportunities such as the reforestation measures in the community of Dottikon and the soil improvements in the community of Boswil.

4.2 *Political acceptance*

Political acceptance is the fundamental basis of every project. It has to be achieved at different levels: First communal authority has to agree. Then the majority off the community members need to be in favour. At last, depending on the project size, the cantonal parliament needs to give its approval.

At the Bünz people's need for flood prevention after the floods of 1994 and 1999 and for local recreation and ecological value in settled areas furthered their political willingness to take action. Restoration became a political programme. A great deal of convincing communal authorities and people was done also by the cantonal agency. This demands a high level of commitment of the cantonal project manager and the maintenance of personal contact to build trust.

Local opinion leaders play an important role in shaping the political acceptance of community members. Working groups help to involve and integrate local stakeholders and thus increase the acceptance (Hostmann et al. 2005). In Wohlen, good solutions for real replacement could be found in the working groups.

Important for acceptance by the municipal council and by the members of the community was also the role of already implemented restorations. They served as a positive showcase. Excursions to already restored sections were organised by the cantonal agency.

Fishery associations and nature conversation organisations like the WWF help strengthen the appreciation of restoration projects and support the authorities by realising them. In the community of Muri, the WWF campaigned for the restoration project and thereby improved the popularity of the project. Furthermore, a water playground was implemented in collaboration with an institution for social community (Hostmann & Knutti 2009).

If the community members and municipal council are persuaded of the project, there is a good chance that the funding will be approved by the parliament.

4.3 *Financing*

In the Canton of Aargau, the Canton is responsible for financing hydraulic engineering projects. The municipalities are obliged to contribute to costs at a ratio of 20% to 60% (§ 116 and § 122 BauG[3]—Cantonal Building Law). The loan to financially secure a project generally has to be approved by the communal council and responsible canton body (governing council or general council). The approving body depends on the loan amount and building contractor.

It was crucial for the financing that synergies with other projects could be exploited. For example, the restoration in Dottikon was able to be financed as part of a compensation for clearing of a forest by an industrial company. In Wohlen, the restoration was carried out to ecologically compensate the flood control reservoir. Flood protection is also the driving force in Waltenschwil, where the project benefits from an increase in government subsidies to 45% thanks to restoration.

Restoration measures without any improvement in flood protection tend to be less well accepted by the public. Subsidies by the federal government enhance political acceptance especially for pure restoration projects. Until 2007 the federal government supported restoration projects in the Canton of Aargau with contributions amounting to 20–25%. With the introduction of financial compensation for environmental areas in 2008 contributions increased to 35%. Since 2011 contributions even up to 80% are possible for revitalisation projects (BAFU 2011).

---

[3]BauG kantonales Baugesetz Gesetz über Raumentwicklung und Bauwesen vom 19.01.1993 (Stand 01.01.2011).

4.4 Land issue

4.4.1 Adequate space for watercourses

Natural bodies of water require more space than artificial, channelled rivers. The federal government issued a recommendation on adequate space for watercourses to ensure flood protection and the ecological functions with the so-called key chart (BWG 2000). The area for watercourse is calculated on the basis of the natural channel bed width and differentiates between the minimum required area for watercourse, biodiversity width and meander belt width. The key chart recommendation was adopted into the Federal Water Protection Ordinance[4] in 2011 (article 41a of the WPO) and is therefore mandatory. The adequate space for watercourses became as well a requirement for subsidies. The wider the area the more the federal government subsidises the project.

4.4.2 Implementation in the Canton of aargau

According to the Cantonal Building Law "...the watercourse's area required to maintain and create the riverbank and plant vegetation along it must be acquired if possible" (BauG § 116). This means that at least the land of the riverbed and—banks has to belong to the Canton. As a dynamic river changes its bed by erosion, additional land needs to be acquired in advance.

The study from 1994 suggests broadening the Bünz plot by 10 m to 30 m (Colenco & creato 1994). The widths of the watercourse plots in Boswil and Dottikon are based on the federal government's key chart. The Canton of Aargau had the watercourse area analysed in a study in 2012 for the Hendschiken, Wohlen and Waltenschwil projects (Flussbau AG 2012). A minimum required area of 25 m was derived based on an analysis of historical maps, plans and morphological criteria with a natural bed width of 7 m. The biodiversity width was defined to be 37 m and the meander belt width 45 m for the Bünz at Hendschiken.

4.4.3 No land no restoration

Exchange for real property is one of the most crucial factors for the political and legal acceptance of a project. In principle, the authority can expropriate the required land for hydraulic engineering projects (§ 132 BauG, article 68 WPA). However, this is the last step and is only applied in rare cases. If possible the land has to be purchased or a solution involving land reallocation must be found (article 11 SSV[5], LwG AG § 10[6]). River restoration therefore depends primarily on large-scale land disposal opportunities. With few exceptions, land compensation could be offered in the Bünz projects. In Boswil—Bünzen the required land could be acquired by the Canton from farmers by land reallocation. In Hendschiken the municipality was able to buy an entire farm which was then used for compensation.

The decisions about the size of the area for watercourses depend on what the land can be exchanged for, the project's objectives and its costs. In Wohlen the Bünz area will be widened to the minimum area required for watercourses. It was not possible to acquire more land, as it was already needed for the flood retention basin. In Waltenschwil a minimum water space was chosen due to agricultural concerns.

5 OUTLOOK ON FUTURE RESTORATIONS

"Adequate space for watercourses, water flow and water quality" these are the guiding objectives for Swiss watercourses (SAEFL/FOWG 2003). Does the Bünz nowadays meet

---

[4]Federal Water Protection Ordinance (WPO) of 28 October 1998 (status as of January 2014).
[5]SVV Verordnung über die Strukturverbesserungen in der Landwirtschaft vom Dezember 1998 (Stand 1. Januar 2014).
[6]LwG AG Landwirtschaftsgesetz des Kantons Aargau vom 13. Dezember 2011 (Stand 1. August 2012).

these objectives? The space for watercourses of the restored sections corresponds at least to the minimum required area (see chapter 4.4.1). The flow of the Bünz can be naturally very low. To reduce impacts during dry periods, water extraction for irrigation purposes is restricted. Due to the renovation of the sewage treatment plants from 1994 to 2009, the pollution level has decreased significantly. In 2009 the power plant Tieffurt-Mühle in Dottikon changed its flushing regime so that the colmation of the river bed which impacted the Bünz in its past could be reduced. For the restored sections the development goals are mostly reached.

A lot has been achieved and more has to be done. I discussed some important aspects for future restoration with project managers of the Bünz. Their inputs are presented below.

Filling the gaps: At the end of 2014, 50% of the 25 km of the Bünz will be restored. The potential for further restoration project with great benefit for nature and landscape is estimated at 5 km. For the implementation of these projects, the funding and the required land need to be provided.

Enhance connectivity: The Jowa weir in Möriken-Wildegg and the Tieffurt-Mühle power plant in Dottikon are two major obstacles to fish migration in the Bünz. They disconnect the Bünz from the important watercourse system Aare and Rhine.

Invasive neophytes: In particular the Japanese Knotweed control is a key issue in the valley of Bünz.

Water course maintenance: The objectives of flood protection and dynamic development of watercourse are partially controversial. Deposits occur more frequent in restored areas. For flood safety they have to be removed, which does not correspond to the aims of a dynamic river. The costs for river maintenance in restored areas tend to increase. The necessary funds have to be provided.

Water quality: The quality of macrozoobenthos, especially in the natural part of the flood plain Bünzauen, has not improved (Lubini et al. 2013) as much as intended. Water quality could restrict the abundance of macrozoobenthos. Urban drainage and intensive agriculture still affect the water quality of the Bünz (Ambio 2009). Several improvement measures are planned.

6 CONCLUSION

The feasibility and the quality of a restoration project considerably depend on the available land and the project context. If land cannot be acquired the restoration work is restricted to existing bed and bank. Widening and variation of the watercourse requires acquisition of land, usually by land replacement. It also depends heavily on the current political context in the municipality. Otherwise there is the risk that too many compromises will be made and the degree of restoration tends to be smaller.

It is important that the planning phase is not drawn out for too long and it can be implemented relatively quickly. In this way reference projects are created that can promote subsequent projects. However, it also needs external promotion. The floods in 1999 in particular boosted the projects.

Splitting into individual projects gave the necessary flexibility in terms of time, planning and construction. Projects could be adapted to the needs of each community and synergies, such as flood protection or land reallocation projects, could be utilised. This approach led to important windows of opportunities. As a result the situation could be reacted to as the occasion arose.

The individual project process does also carry risks. It relies on project managers, who require stamina, alertness, willingness, powers of persuasion and above all perseverance. This is only possible if there is continuity within the responsible department.

9 million SFr. has so far been invested to restore the Bünz; another 6 million SFr. will be spent in near future. As of February 2014 the vision of 50% of 25 km being in a little impacted to seminatural level is within reach. Compared to 1994, when 88% of the Bünz were in a severely impacted or artificial state, it's a success.

ACKNOWLEDGEMENTS

This article was written in collaboration with the Bünz project managers from the agency of landscape and watercourse of the Canton of Aargau. I would like to thank them for their support and information about the Bünz project management. Furthermore, I am grateful to Markus Zumsteg for the discussion about the strategy and the main factors influencing the realisation of the Bünz restoration projects. A special mention goes to Norbert Kräuchi and Erik Olbrecht who reviewed this article.

REFERENCES

Ambio 2009. Erfolgskontrolle an den Gewässern im Einzugsgebiet der Abwasserreinigungsanlagen im Bünztal, ARA-Ausbauprogramm 1994–2009. Ambio GmbH. Zürich.

BAFU 2011. Handbuch Programmvereinbarungen im Umweltbereich. Mitteilung des BAFU als Vollzugsbehörde an Gesuchsteller. Bundesamt für Umwelt (BAFU). Bern. Umwelt-Vollzug Nr. 1105: 222.

BWG 2000. Raum den Fliessgewässern. Bundesamt für Wasser und Geologie (BWG). Bern.

Canton Aargau 2013. Projektgenehmigung Gemeinde Hendschiken. Renaturierung Bünz. Aarau.

Colenco 2002. Moderne Melioration Boswil, Renaturierung Bünz, Ergänzender Bericht zum generellen Projekt. Colenco Power Engineering AG. Baden.

Colenco & Creato 1994. Gewässerstudie Bünztal. Zusammenfassung. Teil A: Einleitung, Synthese, Schlussfolgerungen. Teil C: Gewässerökologische Bewertung. Colenco Power Consulting AG. Baden. creato Netzwerk für kreative Umweltplanung. Ennetbaden.

Flussbau AG 2012. Bünz, Abschnitt Bünzen bis Hendschiken, Herleitung des Gewässerraums. Flussbau AG. Zürich.

Herzog 1918. Bünzkorrektion—Generelles Projekt.

Hostmann & Knutti 2009. Befreite Wasser. Entdeckungsreisen in revitalisierte Flusslandschaften der Schweiz. Rotpunktverlag. Zürich.

Hostmann M., Buchecker M., Ejderyan O., Geiser U., Junker B., Schweizer S., Truffer B. & Zaugg Stern M. 2005. Collective planning of hydraulic engineering projects. Manual for participation and decision support in hydraulic engineering projects. Eawag. WSL. LCH-EPFL. VAW-ETHZ. 48 p.

HZP 2009. Gefahrenkarte Hochwasser Unteres Bünztal, Anhang. Hunziker, Zarn & Partner (HZP). Aarau.

Lubini et al. 2013. Periodische Bestandesaufnahme an grösseren Bächen 2012/2013: Bünz bei Möriken. ARGE Lubini, Vicentini, AquaPlus. 2012/2013.

SAEFL/FOWG 2003. Guiding Principles for Swiss watercourses. Promoting sustainable watercourse management. Bern. 12 pp.

UAG 2005. Die Fischfauna in der Bünz. Die Bünz—Vom Kanal zum dynamischen Gewässer. *Sondernummer 20 aus der Reihe Umwelt Aargau (UAG)*. Kanton Aargau. Aarau.

Swiss Competences in River Engineering and Restoration – Schleiss, Speerli & Pfammatter (Eds)
© 2014 Taylor & Francis Group, London, ISBN 978-1-138-02676-6

Physical modeling of the third stage of Aire River revitalisation project

Z. Vecsernyés & N. Andreini
hepia, Geneva, Switzerland

D. Consuegra
heig-vd, Yverdon, Switzerland

J.-L. Boillat
EPFL, Lausanne, Switzerland

ABSTRACT: The third stage of the Aire River restoration in Geneva is a multi-objective project including a 600 m long dam for flood control, a 8000 m³ sedimentation reservoir and a 1100 m long rehabilitated meandering reach. Beyond river revitalisation issues, the main goal of the project is to optimise appurtenant works dedicated to flood risk management upstream from Geneva City. Due to the high complexity of the project dealing with flood routing, sediment transport, driftwood and fish migration, an optimisation study was carried out on a physical hydraulic model in the Laboratory for Applied Hydraulics of *hepia*-Geneva in parallel with numerical modelling. This paper focuses on choices and improvements related to the original hydraulic design of the project, more particulatrly flow diversion, energy dissipation, driftwood retention, sediment management and fish migration. Significant benefits obtained from the physical modelling are highlighted and illustrated by pictures of the real project under construction.

1 INTRODUCTION

1.1 *Historical context*

In the early XXth Century, a large number of rivers training projects were realized in Europe and most wetlands were drained in order to enlarge the arable land surface and to fight against diseases such as malaria. In 1925 Aire River in Geneva was transformed to a straight-line open channel with a concrete bed and smooth herbaceous embankments. On account of resulting environmental impoverishment, specific articles were included in the Law on water resources of Canton Geneva (LEaux-GE, 1961) with the aim of quality improvement of inland waters as well as restoration and revitalisation of rivers. The project of the Aire River could thus be initiated on the basis of lawful objectives.

The first stage of this revitalisation programme started in 2000 and was dedicated to water quality improvement of the stream, river morphology restoration and rehabilitation of ecological habitats (Gerber 2013). The same year, a flooding of Lully village pointed out the severe lack of the storm water network capacity. The second stage of the project included therefore aspects regarding flood control and river ecology rehabilitation, such as the construction of a storm water detention reservoir, the enlargement of the river bed and the revitalisation of a 2 km long river reach. The third stage of the programme involves a 1.8 km reach with two distinct parts. While the 800 m upstream sector is dedicated for flood detention and free evolution of the mobile river bed, the downstream one spreading over 1.0 km is intended to flood risk management through of hydraulic works.

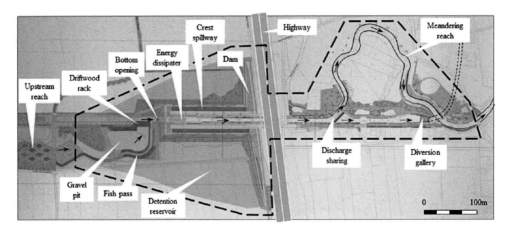

Figure 1. General overview of the Aire River revitalisation project (third stage), with flood risk management hydraulic structures. Dashed line indicates the limit of physical model.

1.2 Main objectives of the project

The main objectives and related flood risk management concepts of the project are:

- Limit the discharge reaching Geneva City to 50 m³/s for a 300 years return period flood by flow retention and diversion.
- Prevent river bed erosion and scouring downstream of the flood control bottom opening by designing an appropriate energy dissipation structure.
- Capture the bed load in a gravel pit on the upstream sector, in order to prevent inundation in Geneva induced by river bed rising during floods.
- Retain driftwood behind a double rack device on the upstream sector, in order to prevent inundation in Geneva induced by bridge sections blocking during floods.

1.3 General description of the project

As presented on Figure 1, the required flood risk management works are:

- A flood detention reservoir extending upstream of a discharge control structure constituted of a bottom opening and a crest spillway.
- A gravel pit.
- A driftwood rack.
- A discharge sharing device, leading the base flow towards the rehabilitated meandering reach and the overflow to an existing diversion gallery.

Since 1D numerical modelling could only provide results on the general hydraulic behaviour of the project, a detailed analysis by means of a small-scale physical modelling revealed necessary for optimisation. Experimental tests concerned not only the hydraulic behaviour but also the alluvium dynamics in the gravel pit, the driftwood transport and retention, as well as the exact behaviour of the control structure elements such as the bottom opening and its energy dissipater, the crest spillway and the discharge sharing work.

2 HYDRAULIC MODEL

2.1 Experimental set-up

The 20 m × 10 m hydraulic model was constructed in the Laboratory for Applied Hydraulics of *hepia*-Geneva, CH, inside the limits indicated in Figure 1. The physical model presented

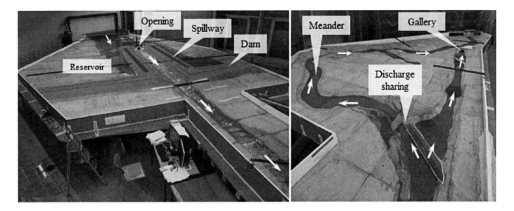

Figure 2. Left: Upper part of the hydraulic model with the dam and the control structure. Right: Downstream part with the discharge sharing work between meander and diversion gallery.

in Figure 2 reproduces a 800 m river reach at a 1:40 length scale ratio. It was operated with respect of Froude and sediment transport similarities. Hydraulic structures such as spillway, energy dissipater, water sharing device and underground diversion gallery were reproduced of PVC and integrated in the mortar shaped topography. Hydraulic structures and topography were modelled with a ±0.5 mm accuracy level on the whole model.

Water discharge was supplied through the inner circuit of the laboratory, by means of three pumps (40 l/s each). Water levels were monitored using 16 ultrasound probes connected via a National Instrument NI 6221 data acquisition module to a computer (accuracy: ±0.5 mm). The upstream inlet discharge was controlled with an Endress-Hauser electromagnetic flow meter (accuracy: ±0.5%). The gallery outlet discharge was measured by a standardised V-notch (accuracy: ±2%). An *ad hoc* LabVIEW software application allowed recording and monitoring simultaneously the water levels and flow rates.

2.2 *Sediment and driftwood characteristics*

The Aire River alluvium was sampled at different locations and the sediment grain size distribution was analysed in the Laboratory for Soil Mechanics of *hepia*-Geneva. The mixture of the non-cohesive sand used on the model was defined in accordance with Shields critical velocity criterion. Shape and size of driftwood were determined with respect of statistics provided by Lange et al. (2006). Total volume of driftwood was determined taking into account the catchment area and recommendations of Rickenmann (1997) and Uchiogi et al. (1996).

3 FLOOD CONTROL HYDRAULIC STRUCTURES

3.1 *Flood control opening and spillway*

The upstream flood risk management structure is made up of a 600 m long labyrinth shaped dam integrating a bottom opening at its centre and two 145 m long spillways on the lateral crests. The opening is characterized by a fixed calibrated geometry limiting the discharge as a function of the upstream hydraulic head. Up to 50 m³/s, corresponding to a 5 year return period flood, no storage occurs (Table 1).

Above 50 m³/s, the opening becomes pressurized, inducing water retention in the reservoir and flood peak reduction downstream of the dam. The reservoir storage volume reaches 300'000 m³ during a 100 year hydrological event, characterised by a 100 m³/s flood peak. Over a 300 year return period flood, the storage capacity of the retention basin becomes insufficient, leading an overflow above the crest spillway. With the perspective to preserve habitations around the reservoir, the required maximum water level is 403 m.a.s.l.. Spillways

Table 1. Discharge upstream and downstream of the control structure and after the discharge sharing work for different return period floods.

T (year)	$Q_{upstream}$ (m³/s)	$Q_{at\ discharge\ sharing}$ (m³/s)	$Q_{meander}$ (m³/s)	$Q_{gallery}$ (m³/s)
<2.33	<20	<20	<20	0
2.33	30	30	24	6
5	50	50	30	20
10	60	58	32	26
30	80	72	35	37
100	100	83	37	46
300	120	100	48	50 (+2 m³/s overflow)

of the dam are therefore positioned at 402.70 m. a.s.l., guaranteeing the overflow and limiting further rise of the water surface.

Numerical modelling preconized a 7.0 m wide and 2.25 m high rectangular flood control bottom opening. Tests on the physical model were carried out for this geometry and further ones. The opening height was first raised to 2.40 m in order to respect the flood routing constraints. Then, respecting an architectural choice, a hexagonal orifice was tested (Fig. 3 left) with an equivalent opening section.

Figure 3 (right) shows the evolution of the reservoir water level as a function of the discharge, respectively for rectangular and hexagonal openings. Up to 50 m³/s, the opening is not submerged. Over 50 m³/s and up to 100 m³/s the bottom opening becomes pressurized, with retention water levels remaining below 402.70 m.a.s.l.. Over 110 m³/s, part of the discharge flows over the lateral spillways, guaranteeing that the water level does not exceed 403 m. a.s.l. even in case of an extreme flood with 150 m³/s peak flow. Since, as shown in Figure 3 (right), rectangular and hexagonal openings gave quasi identical results the latter was finally adopted.

3.2 Energy dissipation device

Energy dissipation downstream of the bottom opening is required to ensure the transition from supercritical to subcritical flow and to protect the downstream natural river reach from scouring and erosion. Since a classical stilling basin is not appropriate to ensure fish migration, the implementation of a macro-roughness structure made of alternate blocks in the river bed offers a good alternative (Fig. 4). In order to determine the appropriate blocks pattern (block shape and spacing, number of rows) as a function of upstream hydraulic head and downstream flow condition, different geomteries and configurations were tested on the model (Fig. 5) as well as in a hydraulic flume (Vecsernyés et al., 2014).

More than 8 configurations with different block shapes and sizes were tested on the model. First, the dissipater was set up with 6 lines composed of plane-parallel sharp-edged blocks. Respecting the architectural choice, cylindrical blocks were then tested. Since cylinders induce a lower drag force than sharp-edged blocks, and in order to achieve comparable energy dissipation performance, two additional block lines were needed. Because of lateral expansion of the flow immediately downstream from the hexagonal orifice, blocks were also added on both channel embankments. At the downstream end of the dissipation structure, a 0.5 m high transversal sill with a 0.10 m deep rectangular notch was added to keep the blocks under submersion and guarantee the fish migration even during low flow conditions. The final configuration of the dissipater is composed of 8 lines containing 6 or 7 alternate cylindrical blocks (diameter 0.8 m, height 0.4 m) respectively in odd and even line numbers.

3.3 Water sharing between meander and diversion gallery

Under low flow conditions (<20 m³/s), the entire amount of water flows towards Geneva via the meander, allowing free fish migration. At moderate discharge and flood conditions, part

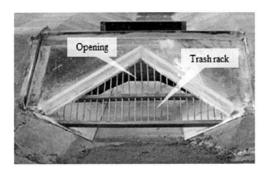
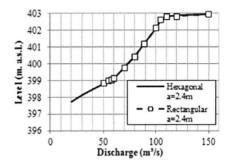

Figure 3. Left: Upstream view of the bottom opening with driftwood trash rack. Right: Water level in the reservoir vs. discharge, under steady state flow conditions (prototype values).

Figure 4. Energy dissipation device. Left: on the model. Right: after construction on prototype.

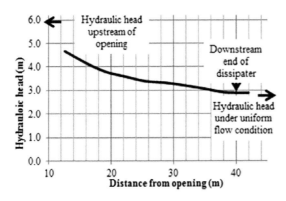
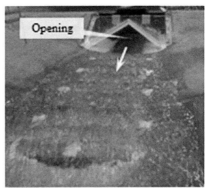

Figure 5. Left: Energy dissipation along the device for Q = 100 m³/s with the final configuration. Right: Dissipater test on model.

of the flow has to be evacuated to the Rhone River by a diversion gallery in order to respect the required limitation for the meander and to save Geneva from inundation.

Numerical simulations revealed that a side weir installed on the right bank at the beginning of the meander could be appropriate to divert the excess discharge toward the existing gallery and thus limit the meander supply to 50 m³/s during a 300 year return period flood.

First tests on the physical model confirmed basic achievements of computations but further analyses were needed to determine the final design of the hydraulic structure (Fig. 6). The sharing device is composed of a side weir, a river restriction and an orifice caused by a bridge.

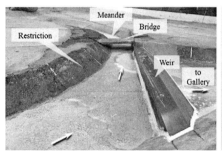

Figure 6.   Left: Final configuration of the discharge sharing work on the model. Right: sharing work after construction on the Aire River.

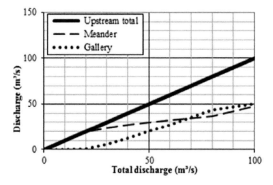

Figure 7.   Discharge sharing between meander and diversion gallery.

Figure 8.   Gallery intake with anti-vortex macro-roughness walls. Left: Top view of the model. Right: Side view on the River Aire.

The ideal water sharing was obtained by successive iterations on weir crest level, channel width and orifice opening under the bridge. As a result, up to 20 m³/s, the total discharge flows in the meander. Above 20 m³/s, a part of flow is diverted over the side weir towards the underground gallery (Table 1). Figure 7 illustrates the progressive discharge sharing in function of the discharge released by the upstream flood control work.

3.4   Underground diversion gallery

The diversion gallery, located downstream of the discharge sharing work, was constructed in 1987 to reduce the flood risk along the urbanized part of the river by leading the overflow to the Rhone River. The hydraulic capacity of the gallery is 50 m³/s.

160

Model tests revealed that when the gallery is submerged at high flow rates ($Q_{gallery} > 45$ m³/s), a vortex forms at the intake, thus reducing the evacuation capacity. In order to prevent the vortex formation, experimental tests were carried out with various damping configurations. The most efficient solution was obtained by putting macro-roughness elements at the channel sides. The final configuration is composed of 4 inclined walls, one in cross-stream direction and three in stream-wise direction (Fig. 8).

4 SEDIMENT AND DRIFTWOOD

4.1 Gravel pit

In order to prevent sediment deposit in the energy dissipation device as well as in the successive pools downstream of it, and to protect Geneva from inundation, sediment has to be captured upstream of the flood control structure. For that purpose, a 8000 m³ gravel pit was created between two sills upstream of the bottom opening. The first sill has a trapezoidal notch to concentrate water during dry weather periods. A breach is fitted out in the second to ensure fish migration.

The hydraulic tests with sediment transport were conducted by increasing the flow rate step by step in order to simulate flood hydrographs. In parallel, the sediment supply was increased gradually at the upstream boundary of the model. After successive flood simulations the basin was completely filled out with an almost flat centred sediment layer (Fig. 9). The amount of released alluvium was not relevant confirming that the gravel pit is efficient.

Figure 9. Sediment deposit (on model) leaving an open channel flow along the downstream end of the gravel pit.

Figure 10. Left: Driftwood stopped by upstream rack, up to T = 30 ans flood. Right: For T > 30 ans driftwood is released from rack and stopped by the trash rack in front of the bottom opening.

161

4.2 Driftwood rack

Driftwood may be a major problem if blocking bridge sections during floods. In order to prevent such risk, floating debris must be captured upstream from the flood control work. Simulations with driftwood were carried out on the physical model by using trunks and root-stocks as described in section 2.2. Different configurations were tested including a trash rack in front of the bottom opening of the control work, a rack on the upstream respectively on the downstream sill of the gravel pit as well as combined solutions.

For each configuration different flood scenarios were tested, with single and multiple consecutive floods. The finally selected configuration consists of a double rack device working as follows: At moderate flow rate, up to 30 year return period peak flow, water level in the detention reservoir is low and driftwood is stopped behind a rack on the downstream sill of the gravel pit (Fig. 10). For higher flow rates, the bottom opening of the flood control work is submerged and the water level rises over the rack, releasing driftwood. The latter is thus stopped like a carpet in front of the trash rack placed above the bottom opening. The driftwood layer doesn't hinder the water to enter into the bottom outlet. In the final configuration the lower part of the opening was left free at 1 m distance from the bottom avoiding accumulation of small debris during low flow conditions (Fig. 3). This solution revealed efficient even after multiple consecutive floods and flow decrease. No blocking of the orifice was observed.

5 FISH MIGRATION

Special attention was paid to guarantee fish migration throughout the whole river reach of the third stage of Aire River project. From downstream, following sectors are concerned:

– The revitalised meander, with its mobile bed and natural spawning ground.
– The sector beween the discharge sharing work and the flood control bottom opening, with successive pools delimited by 40 cm high transversal sills inclouding each a 10 cm notch.
– The energy dissipation cylinders (downstream of the flood control opening), constantly submerged by a 30 cm permanent water depth due to a downstream transversal step.
– An open channel is maintained free along the downstream end of the gravel pit to connect to the fish pass river (Fig. 9).
– The upstream revitalised river reach, with a mobile bed and a natural spawning ground.

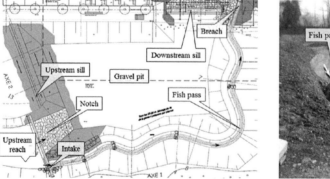

Figure 11. Left: Fish pass project along the right embankment of the gravel pit, and its construction (right).

162

6 CONCLUSIONS

Physical modelling allowed significant improvements of several aspects of the third stage of the Aire River revitalisation project. The most important ones are:

- Optimization of the size and shape of the flood control bottom opening, satisfying hydraulic and architectural requirements.
- Design and validation of an original energy dissipation structure yielding rapid transformation of supercritical to subcritical flow downstream of the opening and preventing the natural river of scouring and erosion
- Design and optimization of the discharge sharing device between the meander and the underground diversion gallery
- Development of an anti-vortex device, maximising the diversion gallery discharge
- Optimization of technical solutions for sediment and driftwood retention
- Control of global and local adequate conditions for fish migration throughout the whole river reach.

REFERENCES

Gerber, F. 2013. Renaturation de l'Aire, agglomération genevoise : une nouvelle façon de créer un cours d'eau. *Ingenieurbiologie Mitteilungsblatt / Génie biologique*. Bulletin n° 1, April 2013. 52–55.

Lange, D., Bezzola, G.R. 2006. Schwemmholz. Probleme und Lösungsansätze. *Mitteilung VAW 188*, Herausgeber H.-E. Minor, ETH-Zürich.

LEaux-GE. 1961. *Loi sur les eaux du canton de Genève.* L 2 05.

Rickenmann, D. 1997. Schwemmholz und Hochwasser; *wasser, energie, luft, 89. Jahrgang.* Heft 5/6: 115–119.

Uchiogi, T., Shima J., Tajima H., Ishikawa Y. 1996. Design methods for wood-debris entrapment. *Internationales Symposium Interpraevent 1996*, Tagungsband, Vol. 5, S. 279–288.

Vecsernyés Z., Destrieux M., Andreini N., Boillat J.-L. 2014. Experimental parametric study of energy dissipater design in natural rivers. *River Flow 2014 IAHR Congress.* Lausanne.

Swiss Competences in River Engineering and Restoration – Schleiss, Speerli & Pfammatter (Eds)
© 2014 Taylor & Francis Group, London, ISBN 978-1-138-02676-6

Restoration of the Broye delta into the Lake of Morat (Salavaux, Switzerland)

E. Bollaert
AquaVision Engineering Sàrl, Ecublens, Switzerland

J. Duval & L. Maumary
Ecoscan SA, Lausanne, Switzerland

S. André & P. Hohl
GE-EAU, Lausanne, Switzerland

ABSTRACT: The Broye River is one of the primary rivers of the Canton de Vaud district. It has been canalized in the 20th century to cultivate the lowlands and to cope with the huge risk of severe inundations of these lowlands. Within this framework, the delta of the Broye River into the Lake of Morat has been canalized and displaced to the north. Since then, several major problems arised. Suspended sediments transported by the Broye River are deposited in a flow recirculation area, resulting in the formation of a sandbank and continuous increase of vegetated surfaces. This resulted in a progressive decrease of the dynamics of the area between the river side and the delta and, finally, also of the delta itself.

Within the framework of a global Swiss river restoration program, the delta of the Broye River will be displaced towards its original, natural course. The aim is to recreate a dynamic alluvial area with regular morphological changes generated by the Broye River and by wave impact during storm events. Second, the new delta gives the opportunity to eradicate invasive plant species, such as the Solidago Canadensis. Furthermore, the project will offer a vegetated area that is fully isolated from anthropic activities and that constitutes a clear separation between natural and built environments, by displacing both hiking trails and mooring places that are actually located along the right bank of the Broye River.

The future morphology of the delta has been predicted by 2D hydraulic-morphologic numerical simulations. These simulations define the long-term morphodynamic potential of the site as a function of its geometry after the works, as well as the influence of for example flow deviating structures. Also, sediment transport of the Broye River as well as storm waves impacting the future delta area have been investigated by means of appropriate numerical modeling.

The present paper summarizes the history of the delta as well as the major natural, technical, morphological and human constraints for its future restoration. The paper points out the role of 2D hydraulic-morphologic numerical modeling in defining the final concept of restoration.

1 INTRODUCTION

The Broye River is one of the primary rivers of the Canton de Vaud district. It has been canalized in the 20th century to cultivate the lowlands and to cope with the huge risk of severe inundations of these lowlands, mainly between Moudon and the Lake of Morat. Within this framework, the delta of the Broye River into the Lake of Morat has been displaced to the north to allow a straight entering of the new channel into the lake.

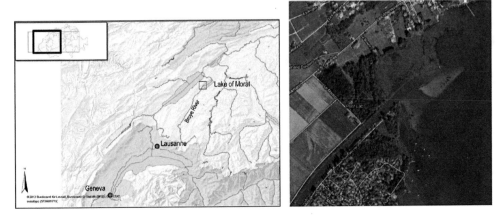

Figure 1. Broye River and localization of its current delta, including sediment deposits (right bank) and immobile reed bed areas (left bank).

However, following these canalization works, several major problems have appeared around the delta. Suspended sediments transported by the Broye River are deposited in a flow recirculation area, resulting in the formation of a sandbank and continuous increase of vegetated surfaces. The sandbank was an important staging area for migrating herons and waders (plovers, stints, sandpipers) and a breeding site for the Common Tern (*Sterna hirundo*). It emerged in 2007 following important floods but disappeared in 2009 under the action of north winds. The deposition of sand continues nevertheless, progressing towards a harbor on the north shore of the lake and is therefore incompatible with navigation. This increasing lack of alluvial dynamics compromises the sustainability of endangered species and plants.

The south shore of the Lake of Morat is a site of national importance according to the Law for waterbirds and migratory birds (OROEM). The mouth of the Broye River is also an important alluvial site (IZA). The shallow waters, rich in food resources, host a number of ducks, herons, waders, gulls and terns. The reed beds are also home for numerous marsh birds, with a colony of several hundreds of Great Crested Grebes (*Podiceps cristatus*).

Within the framework of a global Swiss river restoration program, the delta of the Broye River will be displaced towards its original, naturel course. The aim is to recreate a dynamic alluvial area with regular morphological changes generated by floods and sediments transported by the Broye River and by wave impact during storm events on the lake. Second, the new delta gives the opportunity to eradicate invasive plant species, such as the Solidago Canadensis. Furthermore, the project will offer a more isolated area from anthropic activties and a clear separation between natural and built environments, by displacing the numerous mooring places that are actually located along the right bank of the Broye River. Finally, leisure activities as well as pedestrian access roads will be developed, underlining the sustainable character of the project.

The future morphology of the delta has been predicted by a series of 2D hydraulic-morphologic numerical simulations. In the following, the history of the delta as well as the major natural, technical, morphological and human constraints for its future restoration are briefly described. The role of 2D hydraulic-morphologic numerical modeling in defining the final concept of delta restoration is being assessed.

2 HISTORY OF THE DELTA OF THE BROYE RIVER

The history of the delta of the Broye River is illustrated in Figure 2. The initial natural state of the delta is shown in 1890. The last 300 m of the Broye River is deviating towards the East

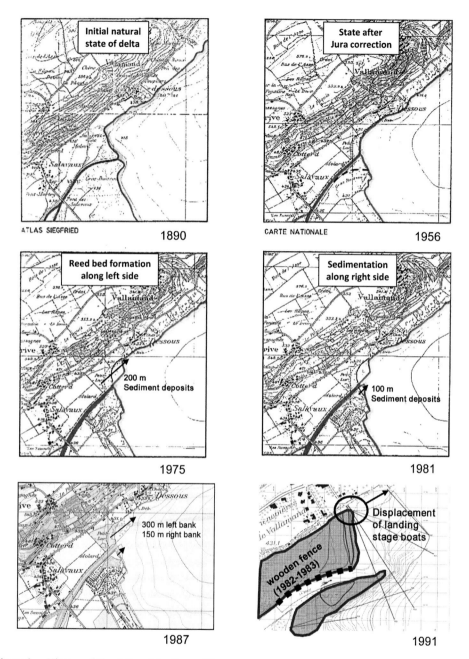

Figure 2.   History of the delta of the Broye River.

to form a naturally growing delta into the Lake of Morat. Between the first (1868–1878) and the second (1962–1973) correction of the Jura Lakes, amongst the different hydraulic works executed at that time, the initial course has been canalized and straightened.

In 1956, a canalization and straightening displaces the delta towards the north. In 1975, sedimentation advances along the left hand side of the delta. As such, over a length of about 200 m, an originally swampy area is progressively being colonized by shrubs and trees. In 1981, progression of the sedimentation along the left hand side results in the right hand side also to colonize, over a length of about 100 m.

167

Figure 3.   a) Common Tern on the sandbank; b) Colonization of the sandbank.

In 1982–1983, the landing stage of the local boats had to be displaced towards the north, and a wooden fence has been constructed along the left hand side in order to deviate the sediments towards the right hand side. This has accelerated the sedimentation along the right hand side.

In 1987, both left and right hand sides of the delta have been extended by 300 m respectively 150 m into the lake and are colonized by shrubs and trees. Within the framework of a highway construction project, about 350,000 m³ of sediments have been extracted from the area.

In 1993–1995, again about 220,000 m³ of fine sediments have been extracted, in order to avoid excessive sedimentation over a period of min. 25 years, i.e. until 2020.

Nevertheless, since 2007, with a draft less than 1 m, navigation is strongly affected by the sedimentation, mainly along the right hand side of the delta. In 2007 a sand bank emerged, quickly colonized by many migrating and breeding birds (herons, waders, gulls and terns) (Fig. 3). It disappeared in 2009 under the action of northern winds. This quick colonization showed the interest of such a sandbank for migrating birds.

3   MAIN PHYSICAL PHENOMENA INVOLVED IN DELTA DEVELOPMENT

The main phenomena involved are illustrated in Figure 4.

3.1   Sediment transport of the Broye River

Based on computations and bathymetries, the Broye River has a mean annual sediment transport of about 20,000 m³, mainly by suspension of sand particles with diameters ranging from 0.1 to 1.0 mm. Suspended transport is occurring on the average for about 18 days per year. Deposition of these sediments when entering the lake forms the basis of the delta progression.

3.2   Delta erosion by wave impact

The head of the delta that progresses into the lake is regularly decapitated by the impact of waves during storm events. With a fetch of about 8 km and winds blowing at about 10–15 m/s during storm events, waves of more than 1 m of amplitude are being generated at the delta for about 12h per year on the average.

3.3   Littoral transport by wave impact

As shown in Figure 5, breaking waves progressively decapitate the delta and generate littoral transport of sediments. Mean annual transport is estimated between 20 and 60 m³/m,

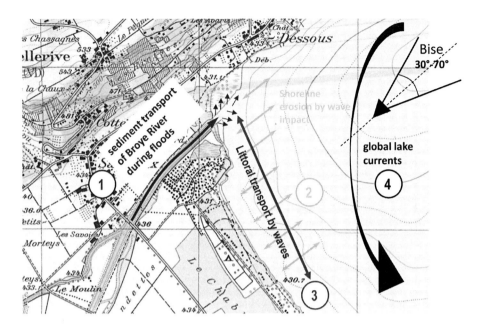

Figure 4.   Main phenomena involved in the development of the delta of the Broye River.

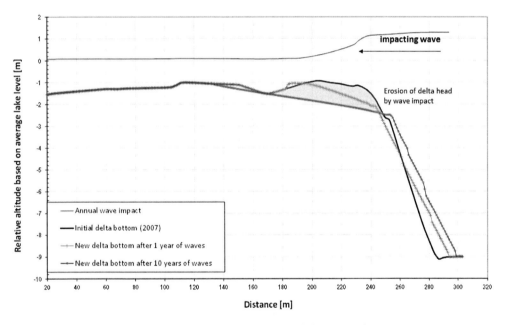

Figure 5.   Erosion by wave impact of the head of the delta of the Broye River.

corresponding to an annual volume of 200–2000 m³ of sediments transported along the lakeshore.

3.4   *Global lake currents*

Global lake currents are oriented towards the south along the shoreline. The importance of these currents remains nevertheless minor compared to the strong littoral currents generated by the breaking waves during storm events.

169

4 NUMERICAL MODELLING OF FUTURE DELTA PROGRESSION

4.1 *Determination of future configuration of the delta*

Detailed 2D numerical modelling (MIKE-21C software) has been performed of both the delta progression by sediment transport and the delta erosion by wave impact. In a first stage, reconstitution of the delta history between 1995 and 2012 has allowed to soundly calibrate the numerical models. Second, different options have been numerically computed and compared with the actual delta situation regarding the following aspects: location of sediment deposits, swampy and alluvial characteristics, potential for environmental improvements, sustainability and costs.

The tested configurations were: regular dredging of sediment deposits, prolongation of the actual wooden fence further into the lake, displacement of the channel back to its original location, and finally a status quo option with only a displacement of the ship landing stage further away and complete acceptance of the environmental degradation.

The most interesting option revealed to be a return to the initial natural situation of the delta, dating from the 19th century, by displacing the last 300 m of the Broye River back into its original riverbed, combined with a lowering of the whole area situated between the actual and the future Broye main channel. As shown in Figures 6 and 7, this will not only allow to redirect future deposits away from the actual deposits and from the ship trajectories, but will also create a morphologically active area, with regular inundations and erosion and deposition of sediments, rather than an inactive zone prone to static reed bed formation.

4.2 *Long-term evolution of future configuration of the delta*

Figure 7 compares the initial state of the delta immediately after the planned deviation with the projected state after 50 years of morphological development of the area. The result is obtained by a long-term 2D numerical modelling of the hydraulics and the morphology

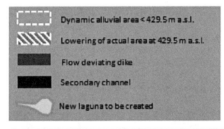

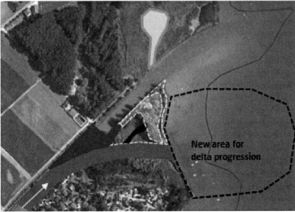

Figure 6.   Displacement of the actual Broye channel towards its original location and development of an alluvial and lakeside dynamic area.

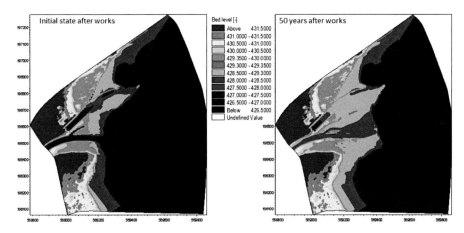

Figure 7.   Erosion by wave impact of the head of the delta of the Broye River.

Figure 8.   a) Greater Ringed Plovor and Dunlins. b) Harvest Mouse nest.

based on sediment transport in the Broye river, but without accounting for the erosional influence of the waves during storm events.

A highly dynamic alluvial and lakeside area is being created between the actual and the future Broye channel, including both inundated and exondated zones that may be modified after each flood event. Beside a partial filling up of the current channel, future sedimentation will be oriented towards the lake and not towards the ship landing stage to the north. Existing mooring points along the actual Broye channel will be displaced and leisure activities as well as pedestrian access roads will be developed. A high quality natural area will be protected and isolated from human activities, underlining the sustainable character of the project.

5   ENVIRONMENTAL ISSUES

The strategic position of the Lake of Morat on the main migration flyway of the Swiss Plateau makes the renaturation of the Broye delta a priority, whilst controlling the sedimentation at the mouth of the river. The environmental issues are the following:

– Flooding of the old arm of the Broye River, especially for the Beaver (*Castor fiber*), the Kingfisher (*Alcedo atthis*) and the Penduline Tit (*Remiz pendulinus*).
– Creation of a secondary channel in the reed bed on the north shore of the river for the herons, the Little Bittern (*Ixobrychus minutus*), the Great Reed Warbler (*Acrocephalus arundinaceus*), the Savi's Warbler (*Locustella luscinioides*), the Reed Bunting (*Emberiza schoeniclus*) and, putatively, for the European Pond Turtle (*Emys orbicularis*). The conservation of the reed beds against the progression of the forest will also favor the Harvest Mouse (*Micromys minutus*) (Fig. 8).

171

a.) actual situation   b.) potential situation on the long term

Figure 9.   Comparison of actual situation and potential future situation of the Broye delta, based on 2D numerical computations of suspended sediment transport in the Broye River.

– Creation of sandbanks facing the river mouth as staging sites for herons, waders (plovers, stints, sandpipers) and gulls and breeding sites for Common Tern (*Sterna hirundo*).

The diversification of habitats will make the delta of the Broye River more attractive for numerous other bird species linked to water.

6   CONCLUSIONS

Based on detailed 1D and 2D numerical modelling of hydraulics and morphology of the delta of the Broye River, long-term future delta development by suspended sediment transport has been estimated for different geometrical configurations of the area. This has allowed optimizing the restoration of the Broye River over its last 300 m before entering the Lake of Morat.

The numerical results illustrate that a highly dynamic and active morphological area may be created (Fig. 9) that will be influenced by both sediment transport during floods in the Broye River and headcut erosion of the delta by wave impact during storm events on the Lake of Morat. The sand banks that will result from the delta progression respect navigation and upstream flood risks, and will offer new staging and breeding sites for the migrating birds (herons, waders, gulls and terns). The creation of a secondary channel in the reed beds on the north side of the river will offer habitat for numerous marsh birds. Also, the new delta will allow reducing invasive species, such as the Solidago canadensis. Finally, leisure activities as well as pedestrian access roads will be developed, underlining the sustainable character of the project.

By displacing the numerous mooring places that are actually located along the right bank of the Broye River, the project offers a clear separation between natural and built environments.

REFERENCES

AquaVision Engineering. 2009. Embouchure de la Broye à Salavaux—Etude des mécanismes d'ensablement des rives du lac de Morat—Etude préliminaire, *Internal Report*.
AquaVision Engineering. 2011. Embouchure de la Broye à Salavaux—Etude morphologique 2D de l'ensablement sur le long terme et analyse des variantes d'aménagement, *Internal Report*.
AquaVision engineering. 2012. Embouchure de la Broye à Salavaux, Etude morphologique 2D de variantes d'avant-projet d'aménagement de l'embouchure*, Internal Report*.
ECOSCAN. 2010. Embouchure de la Broye—Analyse des variantes pour la gestion de l'ensablement—Rapport d'impact sur l'environnement, *Internal Report*.

Swiss Competences in River Engineering and Restoration – Schleiss, Speerli & Pfammatter (Eds)
© 2014 Taylor & Francis Group, London, ISBN 978-1-138-02676-6

Hydropeaking and fish migration—consequences and possible mitigation measures at the Schiffenen Dam

D. Brunner
Lombardi Ltd., Consulting Engineers, Fribourg, Switzerland

B. Rey
Groupe E Ltd., Fribourg, Switzerland

ABSTRACT: This paper present ecological remediation measures at Schiffenen dam in Switzerland, for hydropeaking with a peak flow of 135 m³/s during turbine operation and a residual flow of 5 m³/s. The remediation aims to reduce short-term artificial changes in the water flow, to reduce velocity of water level changes in the river and to improve the ecological aspects of the downstream section. For the fish migration, the proposed and discussed solutions are: migration canal, bypass stream, fish lift and fish pass. These solutions are discussed based on hydraulic, economic, environmental and restoration considerations. Rehabilitation processes such as bed enlargement, gravel bank establishment or wildlife protection are proposed jointly to the constructive solutions.

1 INTRODUCTION

In the river section between the Schiffenen dam and the confluence with the Aare River, the Sarine River suffers from high hydropeaking effects due to power plant operations. The variation of the residual flow affects the hydrology and ecology of the river.

A new federal water protection law will be enforced in 2014 in Switzerland and cantonal authorities have to prepare a hydropower management strategy and to identify the environmental effects of a modification of the hydrological cycle. For the particular case of the Schiffenen dam, several studies have been conducted during the last years treating its exploitation and the downstream effects. The present paper gives an overview and update of these studies regarding the new legislation. The bed load budget is not treated here.

2 THE SARINE RIVER BETWEEN THE SCHIFFENEN DAM AND LAUPEN

2.1 *Results of the hydropower operation*

The natural morphology and alluvial dynamics of the Sarine River have been modified by the construction of the Schiffenen dam in 1964 (Fig. 1). The residual flow is returned to the Sarine by a concreted channel downstream at the dam, which results in high aquatic disturbances. The hydroelectric operation of the plant causes serious harm to the indigenous flora and wildlife as well as their habitats.

The ratio between the maximum flow during turbine operation and the residual flow (Q_{max}/Q_{min}) is reported by the Canton of Fribourg (Etat de Fribourg, Service des ponts et chausses, Section lacs et cours d'eau 2013) to be 18.8/1. This value is high and disrupts the air/water ecotone and is emphasized by the channel with steep bank slopes (2:1) and prevents diverse and layered vegetation.

Figure 1. The Schiffenen dam in Switzerland.

The operation causes short-term artificial changes of flow velocity and water levels. These cycles have disastrous consequences on the aquatic wildlife and the downstream section after Laupen is especially at risk.

2.2 *Legal framework, planning and solutions*

In order to apply the new Federal Act on the Protection of Waters ("LEaux"), the federal government requires mitigation measures at the hydropower plant in three steps: hydropeaking (Baumann, Kirchhofer, Schälchli. 2012), fish migration (Könitzer, Zaugg, Wagner, Pedroli, Mathys. 2012) and bed load budget (Schälchli, Kirchhofer. 2012).

The flows in the diversion channel fluctuate by a factor of 18.8/1. Mitigation measures to attenuate the effect of hydropeaking are therefore required.

The Schiffenen dam, with a height of 47 m, is an insurmountable obstacle for fish and will most certainly be subjected to a fish migration remediation in future. Due to the lack of knowledge concerning downstream fish migration, this decision will be delayed.

The bed load budget is interrupted by the Schiffenen Dam where the sediments settle and are deposited.

3 HYDROPEAKING REMEDIATION

3.1 *Aims*

Based on the studies already conducted, some alternatives have been evaluated in order to improve the ecological conditions by reducing the ratio between the two flow regimes from 18.8/1 to 5/1 and by limiting the speed of artificial fluctuations of the water level from 260 cm/h (4.3 cm/min) to 12 cm/h (0.2 cm/min). These alternatives are discussed below.

3.2 *Alternative 1: "Water diversion without electricity production"*

To reduce the hydropeaking effects downstream of the Schiffenen dam, the diversion of 130 m³/s or 110 m³/s from the toe of the dam to the Morat Lake (Fig. 2) is studied. A ratio between the maximum hydropeaking flow and the minimum flow of 1/1 and 1/5 respectively, can be assumed (Lombardi Ltd. 2013). The preliminary design foresees the following structures: an intake in the tailrace canal downstream of the dam and an excavated penstock. A gentle slope of 0.5% allows a water transfer over 10 km and a net head of 50 m until the Morat Lake. This diversion will cause an elevation, a drawdown of a few centimeters of the Morat Lake and a loss of production for the hydropower plants located downstream, namely: Niederried, Kallnach, Aarberg and Hagneck which are located on the Aare River downstream Laupen. The construction costs are estimated at CHF 160 million.

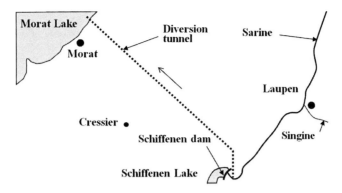

Figure 2. Alternative 1: "Water diversion without electricity production".

3.3 Alternative 2: "Water diversion including power production (turbine or pumped storage generation)"

The second alternative (Fig. 3) aims to reduce the hydropeaking effects on the Sarine River downstream the Schiffenen Dam. Ensuring a 5 m³/s constant residual flow, the idea is to use the net head of about 100 m between the Schiffenen Lake and the Morat Lake for electricity production, either with a new hydroelectric plant in pumped-turbine or only turbine mode. For that purpose a new plant will be built including an intake, a headrace tunnel of 6.9 km, a surge tank, a penstock, an underground powerhouse, a 1.8 km tailrace and an intake structure in Morat in case of pumped water storage. The additional annual production is estimated at 140 GWh. The environmental consequences will be water level fluctuations in the Morat Lake and water mixing from various catchment areas (Broye and Sarine). The construction costs are estimated at CHF 350 million for the turbine and CHF 450 million for the pumped-storage scheme.

3.4 Alternative 3: "Retention"

The alternative "Retention" (Fig. 4) is designed to provide a retention volume partially excavated downstream of the dam. The retention volume of about 3.6 million m³ is based on the actual directive (Schälchli, Kirchhofer. 2012) and allows a peak flow reduction from 135 to 87 m³/s and an increase of the residual flow from 5 to 15 m³/s and fulfills the recommended 5/1 ratio (Meile, Fette. et al. 2005).

The actual mode of operation and current use of turbines only allows a reduction of water level rise from 5.68 cm/min to 1.14 cm/min and is therefore closer to the recommended value of 0.2 cm/min (Meile, Fette. et al. 2005).

The effectiveness of the retention volume, inspired by the concept of a multipurpose reservoir (Heller, Schleiss. 2011), can be optimized significantly by taking into account the meteorological forecast for the lake management, the use of the residual water flow for electricity production or the setting up of a pumped-storage between the restoration basin and the Schiffenen Dam. The construction costs of the alternative "Retention" are estimated at CHF 500 million.

3.5 Alternative 4: "Channel widening"

In order to reduce hydropeaking effects, the concept of the alternative "Channel widening" (Fig. 5) uses the same structures as alternative 3. The construction of an embankment is planed near the confluence of the Sarine River and the Singine River at Laupen. The retention volume of 3.6 million m³ will be excavated into the banks of the Sarine River. Fish migration will be possible by a fish ladder or a lift. The construction costs are estimated at CHF 650 million.

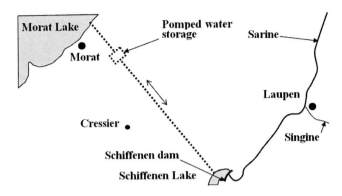

Figure 3.   Alternative 2: "Water diversion including power production (turbine or pumped-storage mode)".

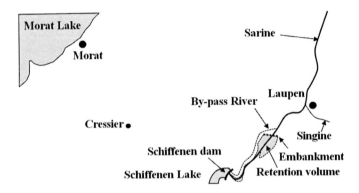

Figure 4.   Alternative 3: "Retention".

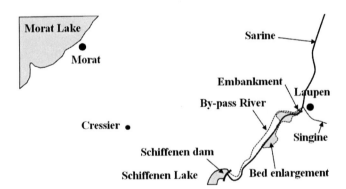

Figure 5.   Alternative 4: "Channel widening".

3.6   *Alternative 5: "Modification of plant operation"*

To reduce the impacts of hydropeaking, the modification of plant operation may be considered. According to current law (art. 39a, al. 1, LEaux), operation modification can be taken on a voluntary basis by the hydropower plant operator. The cost of plant operation modification is typically about 3.5 times higher than other construction solutions. (Wickenhäuser. et al. 2004).

4 FISH MIGRATION IMPROVEMENT

4.1 *Goals*

For the particular case of the Schiffenen dam, fish migration improvement should ensure a connection between downstream and upstream reaches of the dam. Different alternatives have been analyzed and are presented in the following chapters.

Affected fish species are brown trout, lake trout, grayling, nase, catfish and salmon. The latter was found in the Sarine River before dam construction on the Rhine River. Brown trout migration season is between September and February, grayling and nase between March and May and the one of catfish between May and July.

4.2 *Alternative migration channel*

If alternatives 1 or 2 "Water diversion" are applied, the concrete channel downstream of the dam could be used for the fish migration. Actually, the salmonids can already reach the bottom of the dam but with the reduction of hydropeaking, it would be also possible for the cyprinids. Technical alternatives to reach the reservoir have to be discussed (fish ladder; fish lift). This alternative only sets the target for restoring fish migration between the Sarine River in Laupen and the Schiffenen Lake. It does not take into account the requirement of the legislature to improve the habitats necessary for the development of the species.

4.3 *By-pass river*

A semi-natural by-pass river between the Sarine River, just before its confluence with the Singine River and the Schiffenen Lake would have been a perfect solution for the fish migration. However, the level difference between the lake and the agricultural plain downstream of the dam is 38 m and the topographic configuration does not allow the construction of a by-pass river in case of alternatives 1 and 2.

In case of alternatives 3 and 4, a by-pass river followed by a fish lift would be possible (Fig. 3 and Fig. 4). A fish ramp with dispersed blocks would also be required to join the Sarine bed just downstream of the embankment. The cost of a 2.5 km semi-natural by-pass can be estimated at CHF 5 million.

4.4 *Fish lift*

Fish migration is ensured by a fish lift (Fig. 5), which attracts migratory fish species. It is a mechanical structure that catches migratory fish species downstream and then lifts them up to the Schiffenen Lake. A system of water supply as well as an energy dissipation ladder is also needed. The flow in the fish lift is controlled and adjusted hourly, daily and seasonally depending on the frequency of migratory fish species. Fish migration is monitored using a digital camera. The cost of implementing a fish lift is approximately CHF 2 million.

4.5 *Fish ladder*

In order to restore fish migration in case of alternatives 1 and 2, a specific fish ladder should be constructed at the bottom of the dam (Fig. 6). The entrance of the fish ladder will be located near the outflow from the dam.

The basin volume should be at least 5 m³ to allow sufficient energy dissipation (150 watts/m³) and the head between the basins should not exceed 15 cm in order to match the tolerances of the cyprinids (Hefti. 2012).

4.6 *Morphological restoration of the Sarine River downstream the Schiffenen Dam*

According to the natural process of bed evolution, the Sarine would not have eroded as deeply in a trapezoidal channel with 5 to 15 meters depth.

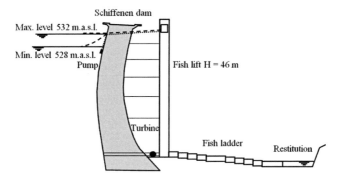

Figure 6. Combination of fish lift and fish pass in case of the alternatives 1 and 2.

Table 1. Comparison of results of mitigation measures.

Targets/alternatives	1	2	3	4	5
Society: flood protection & recreation	2	2	0	−1	4
Environment	15	15	10	10	9
Economy	3	4	−1	−2	−5
Total	20	21	9	7	8

As a result, some river bed widenings at specific river sections may be interesting in the context of the diversification of bed morphology and creation of fish hiding refuges. These enlargements could be arranged and designed in order to encourage the connection of small streams sideway. However, these extensions require surfaces and significant earthworks.

In the context of rehabilitation, the restoration of gravel banks would also improve bed morphology and restore habitats historically present in this section. However, this can only be realized in combination with alternatives 1 and 2 "Water diversion".

5 MULTI-CRITERIA COMPARISON OF ALTERNATIVES

A multi-criteria comparison of alternatives considering fish migration has been made using the targets of sustainable development: social, environmental and economic aspects (Woolsey. et al. 2005). Table 1 gives the score of the topics and subtopics such as obtained by adding up of the following: − − − (worse case), 0 (zero state) and +++ (best case).

By transferring hydropeaking discharges directly towards the Morat Lake, alternatives 1 and 2 offer also a flood security using the retention capacity of the Jura Lakes. Potential ecological impacts are not well known yet. A rehabilitation of the river reach between the Schiffenen Dam and Laupen and also technical improvements at the toe of the dam will be necessary to improve the ecological value of the current channel and to ensure fish migration. In terms of economic effectiveness the diversion of water in combination with a hydropower plant is the most interesting alternative, because it offers, besides hydropeaking remediation, an increase of renewable and local energy production.

6 CONCLUSION

In view of hydropeaking and fish migration remediation as well as sustainable energy production, alternative 2 "Water diversion including electricity production" shows the best cost—benefit ratio. However it requires significant technical measures for fish migration and a revitalization of the Sarine channel downstream the dam. Due to an increase of hydroelectrical production complying with environmental regulations, alternative 2 has a better global efficiency regarding social, environmental an economic aspects.

REFERENCES

Baumann P., Kirchhofer A., Schälchli U. 2012. Assainissement des éclusées—Planification stratégique. Un module de l'aide à l'exécution Renaturation des eaux. Office fédéral de l'environnement, Berne. L'environnement pratique n° 1203: 127 p.

Etat de Fribourg, Service des ponts et chausses, Section lacs et cours d'eau. 2013. Planification stratégique de l'assainissement des éclusées—rapport intermédiaire provisoire. Fribourg: BG Ingénieurs conseils et PRONAT SA.

Heller Ph., Schleiss A., 2011. Aménagements hydroélectriques fluviaux à buts multiples: résolution du marnage artificiel et conséquences sur les objectifs écologiques, énergétiques et sociaux, La Houille Blanche No. 3, p. 34–41.

Könitzer C., Zaugg C., Wagner T., Pedroli J.C., Mathys L. 2012. Rétablissement de la migration du poisson. Planification stratégique. Un module de l'aide à l'exécution. Renaturation des eaux. Office fédéral de l'environnement, Berne. L'environnement pratique n° 54 S.

Lombardi Ltd. 2013. Barrage de Schiffenen. Déviation des eaux dans le lac de Morat. Etude de faisabilité, Freiburg.

Meile T., Fette M., et al. 2005. Synthesebericht Schwall/Sunk. Publikation des Rhone-Thur Projekts. Eawag, Kastanienbaum.

Woolsey S., Weber C., Gonser T., Hoehn E., Hostmann M., Junker B., Roulier C., Schweizer S., Tiegs S., Tockner K. & Peter A. 2005. Guide du suivi des projets de revitalisation fluviale. Publication du projet Rhône-Thur. Eawag, WSL, LCH-EPFL, VAW-ETHZ, 113 p.

Wickenhäuser M., Hauenstein W., Minor H.-E. 2004. Schwallreduktion bzw. Hochwasserspitzenminderung im Alpenrhein, Bericht im Auftrag der IRKA, Projektgruppe Energie.

Swiss Competences in River Engineering and Restoration – Schleiss, Speerli & Pfammatter (Eds)
© 2014 Taylor & Francis Group, London, ISBN 978-1-138-02676-6

Flow restoration in Alpine streams affected by hydropower operations—a case study for a compensation basin

M. Bieri
Pöyry Switzerland Ltd., Zurich, Switzerland (Formerly Ecole Polytechnique Fédérale de Lausanne, Lausanne, Switzerland)

M. Müller
IUB Engineering Ltd., Berne, Switzerland

S. Schweizer
Kraftwerke Oberhasli Ltd., Innertkirchen, Switzerland

A.J. Schleiss
Ecole Polytechnique Fédérale de Lausanne, Lausanne, Switzerland

ABSTRACT: Hydropeaking, resulting from rapid starting and shut-down of turbines, is one of the major hydrological alterations in Alpine streams. The upper Aare River basin in Switzerland comprises a complex high-head storage hydropower scheme. The significant turbine capacities of the two downstream powerhouses produce severe hydropeaking in the Aare River. To reduce the negative impact of the foreseen increase of the turbine discharge, a compensation basin combined with an extended tailrace tunnel downstream of the powerhouses has been designed and is under construction now to facilitate lower flow ramping increasing time for aquatic species to react. The design of the basin and its overall operation had to be defined to reach best ecological as well as economic performance. The retention volume and the operation rules of the basin have been optimized to avoid dewatering of juvenile brown trout. Further, flow ramping has to be limited in order to reduce drifting of macroinvertebrates. The paper presents a consistent approach of a target-oriented process management, including modelling, simulation and comparison of future flow regime without and with mitigation measure. Finally, rules for decision-making as well as the prototype's final design are addressed.

1 INTRODUCTION

Since 1950, a large number of high-head storage Hydropower Plants (HPPs) in the Alps have supplied peak load energy to the European power grid (Schleiss 2007). In Switzerland, for example, 32% of the total electricity in 2010 was produced by storage hydropower plants. Water retention in large reservoirs and concentrated turbine operations allow electricity to be produced on demand. The sudden opening and closing of the turbines produces highly unsteady flow in the river downstream of the powerhouse (Moog 1993). This so-called hydropeaking is the major hydrological alteration in Alpine regions (Petts 1984, Poff et al. 1997). Due to the unpredictability and intensity of flow change, sub-daily hydropeaking events disturb the natural discharge regime, a key factor in ecological quality and the natural abiotic structure of ecosystems (Parasiewicz et al. 1998, Bunn and Arthington 2002). These disturbances directly affect riverine biological communities (Young et al. 2011). Frequent and rapid fluctuations change hydraulic parameters, such as flow depth, velocity and bed shear stress (Petts and Amoros 1996), and thus influence habitat availability, stability and quality.

After decades of the extensive use of water resources, with severe consequences for aquatic and riverine biota, governments have begun to recognise the need for a water protection policy, e.g. the European Union Water Framework Directive. In Switzerland, Parliament adopted the Law on Water Protection in 2011 to improve the quality of Swiss waters, including hydropeaking mitigation.

In a first step and according to an upgrading programme of the hydropower scheme of the Kraftwerke Oberhasli (KWO), the flow regime of the upper Aare River in Switzerland should be improved and thus the hydrological deficit reduced. Several studies (Schweizer et al. 2010, 2012, 2013a, b, c, d) have analysed the aquatic habitat conditions regarding the recently published guidelines of the Swiss Confederation (Baumann et al. 2012). Ecological conditions are supposed to get significantly improved by a reduction of up- and down-ramping rates respectively. A comparison of several mitigation alternatives (Person et al. 2014) revealed a compensation volume between the turbine releases of the Innertkirchen 1 and 2 HPPs and the Aare River as ecologically and economically most effective. The goal of the herein presented study is, on the one hand, to define needed retention volume as well as the operation rules by an optimization algorithm and, on the other hand, the detailed layout of the mitigation measure, consisting of a retention basin combined with an extended tailrace tunnel of the Innertkirchen 1 HPP. Simplified conditions had to be re-evaluated for prototype's design and implementation. The following chapters mainly focus on the challenging step from the preliminary modelling to final design of the first hydropeaking retention basin in Switzerland, assessed according to the new hydropeaking guidelines of the Swiss Confederation.

2 CASE STUDY

Figure 1 shows the upper Aare River basin located upstream of Lake Brienz in the centre of the Swiss Alps. The surface area is 554 km², of which about 20% is glaciated. The natural hydrological regime of the Aare River, with a mean annual discharge of 35 m³/s, shows low discharge in winter and high runoff in summer due to snow and glacier melt. The mean catchment altitude is 2150 m a.s.l. The Aare River, also called the Hasliaare at its headwaters, has its source in the Unteraar and Oberaar glaciers (Schweizer et al. 2008).

Since the early 20th century, a hydropower scheme of nine powerhouses and several reservoirs and intakes has been constructed. The Kraftwerke Oberhasli (KWO) Company utilises 60% of the catchment area for hydropower. KWO has a total installed capacity of 650 MW

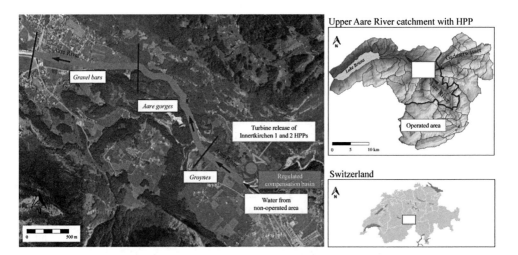

Figure 1. Aare River reach downstream of tailrace tunnels of Innertkirchen 1 and 2 hydropower plants and its location in the upper Aare River catchment upstream of Lake Brienz and in Switzerland.

and generated 1750 GWh (without pump-storage) in 2010, corresponding to approximately 10% of the Swiss hydropower output. The water from the Aare catchment flows through the artificial reservoirs of Oberaar, Grimsel, Räterichsboden and Handeck. In Innertkirchen, the water is returned to the Aare River by the Innertkirchen 1 HPP. The River Gadmerwasser drains the eastern part of the basin (Susten). After driving the turbines, the water is released from the tailrace of Innertkirchen 2 HPP to the Hasliaare River. The substantial turbine capacities of the Innertkirchen 1 and 2 HPPs of 39 and 29 m^3/s respectively produce severe hydropeaking in the downstream river. An upgrading programme for the entire scheme, called KWO*plus*, comprises a large number of technical, economic and ecological improvements to the scheme. To compensate for the turbine capacity increase of Innertkirchen 1 HPP by 25 m^3/s, a compensation basin downstream of the powerhouse outflow is planned for reduction of flow up- and down-ramping rates.

In the 19th century, the dynamic braided river network of the Hasliaare River was drained for agricultural use and flood control. A mostly straight channel resulted from the pristine braided network because of the successive river channelisation. Based on the three parameters of variability of water surface width, bank slope and mesohabitat, the reach downstream of the powerhouse outlets can be divided into four reference morphologies: a reach with artificial groynes (650 m), the Aareschlucht Canyon (1.4 km), a reach with alternating gravel bars (1.3 km) and a monotonous and straight channel reach (11 km). The dewatered reach upstream of Innertkirchen, which carries residual flow, has in its upstream part a natural morphology. Mainly the gravel bars still show natural morphology with varying instream structure.

The condition and type of habitat influence species diversity, growth rates and abundance of aquatic fauna and flora. Several studies have been performed for analysis and understanding of the ecosystem of the Hasliaare River (Schweizer et al. 2010, 2012, 2013a, b). Fish as well as benthos are especially relevant regarding hydropeaking. The quality of the aquatic habitat of the Hasliaare River has decreased during the last 150 years. The dynamic braided river network with various mesohabitats gave way to a mainly straight and monotonous channel without any instream structure. Since the 1930s, the natural flow regime of the river network in the upper Aare River catchment has been altered by high-head storage schemes. Seasonal water transfer from summer to winter and an increased frequency of daily peak discharge events result. Abundance and biomass of fish and benthos have decreased due to the negative influences. Despite today's situation of aquatic biota, the potential for biological development of the Hasliaare River has been highlighted. Investigations to improve the river morphology and the flow regime have been therefore recommended.

3 METHODS

Figure 2 shows the main steps of the procedure of the evaluation and implementation of a hydropeaking mitigation measure. The analysis of the Hasliaare River highlighted beside the morphological deficits a hydrological mitigation potential. Thus, the retention volume between the tailwater of the Innertkirchen 1 and 2 HPPs and the stream, consisting of a basin and a tunnel, should allow (1) minimizing up-ramping and thus macroinvertebrates' drifting and (2) dampening of down-ramping, avoiding dewatering of juvenile brown trout in the gravel bars reach for low flows (<8.1 m^3/s).

4 HYDROLOGICAL SIMULATION

4.1 *Flow up- and down-ramping*

Hydropeaking is mainly critical in winter due to generally low runoff from the catchment area. Thus, the study focused on winter periods between mid-November and mid-March from 2009 to 2012. For the four winter periods, 15-minutes data series of Innertkirchen 1 and 2 turbine release in addition to the runoff from the non-operated catchment have been considered.

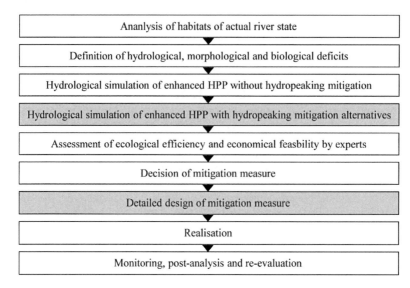

| Ananlysis of habitats of actual river state |
| Definition of hydrological, morphological and biological deficits |
| Hydrological simulation of enhanced HPP without hydropeaking mitigation |
| Hydrological simulation of enhanced HPP with hydropeaking mitigation alternatives |
| Assessment of ecological efficiency and economical feasbility by experts |
| Decision of mitigation measure |
| Detailed design of mitigation measure |
| Realisation |
| Monitoring, post-analysis and re-evaluation |

Figure 2. Methodology of definition, design, realisation and monitoring of hydropeaking mitigation measures as applied for the upper Aare River catchment.

As the turbine capacity of the Innertkirchen 1 HPP of 39 m³/s is increased by 25 m³/s in the framework of the upgrading programme KWO*plus*, data series of future operation had to be generated. When the sum of the Innertkirchen 1 and 2 release was greater than 55 m³/s, full 25 m³/s were added. For values smaller than 35 m³/s, no additional release was considered. In between, proportional addition was applied.

The goal of dampening of flow ramping is not to reduce peak discharge Q_{max} or increase off-peak discharge Q_{min}, but to achieve flow change over a longer time lap and thus to reduce the flow gradients. The flow ramping rate $\Delta Q(t)$ [m³/s/min] indicates the discharge increase or decrease respectively over a given time step, whereas up-ramping generates positive values and down-ramping negative ones:

$$\Delta Q(t) = \frac{Q(t) - Q(t - \Delta t)}{\Delta t} \tag{1}$$

where $Q(t)$ = discharge at moment t; $Q(t-\Delta t)$ = discharge at moment $t-\Delta t$; and Δt = time step.

The flow level ramping rate $\Delta H(t)$ [cm/min] indicates the change of flow level:

$$\Delta H(t) = \frac{H(t) - H(t - \Delta t)}{\Delta t} \tag{2}$$

where $H(t)$ = flow level at moment t; $H(t-\Delta t)$ = flow level at moment $t-\Delta t$; and Δt = time step.

In the given case, up- and down-ramping had to be distinguished, as involved in different biological phenomena. Further, a comparison of the hydrographs immediately downstream of the powerhouses and at the gravel bars reach allowed the definition of the morphology induced damping effect of the corresponding river reaches. Ramping rates of the generated flow series could be correlated to the downstream ones.

Down-ramping is only crucial for the gravel bars reach, as only there dewatering of brown trout is a risk. 2D hydrodynamic simulations allowed the definition of flow level down-ramping rates from the generated flow series for the two cross-sections. As a result dewatering is thus only a problem between 8.1 and 3.1 m³/s. The guidelines (Baumann et al. 2011) define a maximum flow level down-ramping of −0.5 cm/min.

4.2 Operation

Today's as well as future's flow regime should be compared to flow regime influenced by retention volumes. For the parameter study volumes of 50'000, 60'000, 80'000 as well as 100'000 m³ have been modelled, operated and compared. The whole turbine release from the Innertkirchen 1 and 2 HPPs is given to the compensation volume, which has to be operated reducing up- as well as down-ramping and guarantee a minimum discharge of 3.1 m³/s. To reduce down-ramping, water should be retained in the basin and released as slowly as possible. Doing this too slowly, only little volume would be available in case of starting turbines and up-ramping could be reduced less efficiently. Turbine release was just given for one time step of 15 minutes. Two scenarios have been set up and assessed:

– *Scenario A* minimises the up-ramping rate by respecting today's down-ramping rates.
– *Scenario B* optimises firstly the down-ramping for low flows and secondly the up-ramping rate. To guarantee enough retention volume for down-ramping for low flows, a volume of 12'000 m³ is retained.

The comparison between the different alternatives has been undertaken with the 95%-percentile of daily maximum values.

4.3 Results

For today's scheme and for winter conditions of 2009 to 2012, the flow up-ramping rate is 1.36 m³/s/min (Fig. 3a). The flow level down-ramping rate for discharges below 8.1 m³/s of −2.5 cm/min for the gravel bars reach (Fig. 3c) is much higher than the recommended limit value of −0.5 cm/min in the guidelines (Baumann et al. 2012).

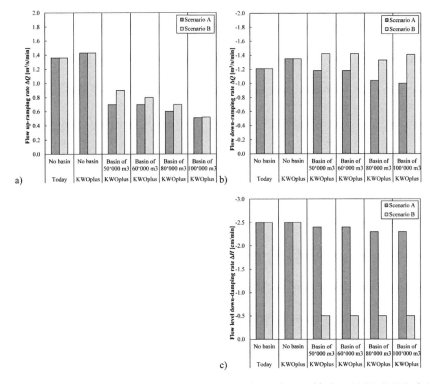

Figure 3. Flow regime characteristics for today's and the enhanced Innertkirchen 1 HPP (KWO*plus*) without and with retention volumes for scenarios A and B for winter conditions from 2009 to 2012: 95%-percentile of (a) flow up-ramping rate and (b) down-ramping rate immediately downstream of the outlet of the retention volume; and (c) flow level down-ramping rate for flows below 8.1 m³/s for the gravel bars reach.

185

Increasing the discharge capacity of Innertkirchen 1 HPP by 25 m³/s, flow up-ramping rates would slightly increase compared to today's values (Fig. 3a). The down-ramping rate as well as the flow level down-ramping value for discharges below 8.1 m³/s remain the same (Fig. 3b and c).

Scenario A shows the reduction ability of flow up-ramping of increasing retention volume, from 1.36 (today) and 1.43 m³/s/min (KWO*plus*) to values of 0.70 and 0.51 m³/s/min for 50'000 and 100'000 m³ retention capacity respectively (Fig. 3a). The down-ramping rates generally decrease (Fig. 3b). The compensation volume allows also to reduce extreme values (100%-percentile). The smallest volume of 50'000 m³ would decrease the flow up-ramping rate from today 2.43 to 1.06 m³/s/min and increase the down-ramping rate from −2.82 to −1.49 m³/s/min.

Scenario B is able to reduce flow level down-ramping for flows below 8.1 m³/s for the gravel bars reach in any case to −0.5 cm/min, achieving the implemented threshold value (Fig. 3c). It reduces flow up-ramping rates to values of 0.90 and 0.52 m³/s/min for 50'000 and 100'000 m³ retention capacity (Fig. 3a). Extreme values are also affected. The order or magnitude is slightly lower than for scenario A, as 12'000 m³ of the volume are used for low flow down-ramping.

Based on the generated data and the construction cost estimates, an expert panel of environmental specialists, engineers, representatives of cantonal and federal authorities as well as the owner assessed the alternatives by a cost-benefit-analysis. Finally, the 80'000 m³ alternative has been selected as the most convenient compromise, acceptable for all of the involved partners.

5 DETAILED DESIGN

The enhancement of the existing Innertkirchen 1 HPP in the framework of KWO*plus* consists of an additional headrace tunnel and powerhouse, called Innertkirchen 1E, which is actually under construction. From the surge tank, a second pressurised shaft guides water toward the new Innertkirchen 1E powerhouse (Fig. 4). The existing tailrace tunnel will be connected to the new one and will be closed by a bulkhead gate at its downstream end. Thus,

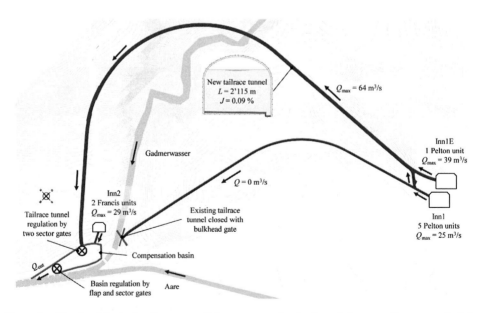

Figure 4. Sketch of the hydraulic system of the compensation basin and the new tailrace tunnel of the enhanced Innertkirchen 1 HPP (Inn 1 and Inn 1E) and the Innertkirchen 2 HPP (Inn 2).

the released water of the enhanced HPP will flow through the new tailrace tunnel into the compensation basin.

Based on the results of the hydrological simulation and in the framework of the realisation of the HPP enhancement, the IUB Engineering Ltd. has studied several alternative concepts for the compensation basin. As the on-site conditions do not allow the construction of a surface basin of the required volume of 80'000 m³, the new 2.1 km-long tailrace tunnel between the Innertkirchen 1E powerhouse and the basin is extended to provide extra volume and to actively contribute to compensation of flow variations from the turbines. Two sector gates at the downstream end of the tailrace tunnel as well as a flap gate and a sector gate at the outlet of the compensation basin allow operation of the two storage volumes.

5.1 Regulation strategy

In addition to the preliminary simulations, which basically considered the ecological requirements and one homogenous volume for flow regulation, the detailed design had to deal with system and operation constraints given by the plant operator KWO. Based on instantaneous $Q_{in}(t_0)$ as well as predicted $Q_{in}(t_1) = Q_{in}(t_0 + 15')$ turbine discharge of Innertkirchen 1, 1E and 2 HPPs, the instantaneous discharge released to the Aare River by the compensation basin $Q_{out}(t_0)$ and the stored volume $V(t_0)$, the regulation algorithm calculates the discharge $Q_{out}(t_1)$ which has to be released during the next time step, taking into account:

– *Priority 1—System reliability and safety*: Released outflow $Q_{out}(t_1)$ would neither empty the basin nor lead to its overflow within the next time step. As the tailrace tunnel is long, routing effect leads to flow propagation time of 7 minutes. When high inflow is predicted, the control gates of the basin have to be opened for preliminary water release in order to ensure the up-ramping rates. However, during the period of pre-up-ramping, the basin itself should not be emptied.
– *Priority 2—Maximum up- and down-ramping rates:* Up-ramping rate is limited to 2.5 m³/s/min and down-ramping rate to −2.5 m³/s/min for discharge higher than 8.1 m³/s and to −0.14 m³/s/min for low flows.
– *Priority 3—Operation flexibility*: The released discharge $Q_{out}(t_1)$ is set, that is within the next two time steps of 15 minutes (t_2) operation can either be stopped to $Q_{in}(t_2) = 0$ m³/s or increased to full capacity $Q_{in}(t_2) = Q_{max} = 93$ m³/s.
– *Priority 4—Desired up- and down-ramping rates*: Up- and down-ramping rates, which are ecologically desired but not crucial for species survival, are taken into account by the regulation whenever possible.

Based on these regulation rules, the retention volume is managed. The boundary conditions define a range of possible released discharges $Q_{out}(t_1)$. Within this range, the regulation algorithm calculates the optimum discharge with respect to the inertia of the system as well as the upper and lower discharge limits, improving the regulation performance.

A regulation at high inertia leads to no or very small discharge variations as long as the operation flexibility and the system safety are not affected. Thus, the retention volumes are exploited to a maximum, as they are filling or emptying until the prioritised boundary conditions become relevant. On a technical level, this results in less but bigger regulation movements of the gates. However, when the system reaches its limits, i.e. the prioritised boundary conditions become relevant, flow changes between two time steps are big and thus up—or down-ramping rates are reaching values close to the maximum of 2.5 m³/s/min. Low inertia requires a higher technical complexity of gate control but leads to faster adaptation of released discharge between two time steps. Thus, the operation and system safety boundary conditions are reached less often as the regulation reacts more rapidly on flow changes from the HPPs.

The upper and lower discharge limits are defined by maximum and minimum turbine discharges forecasted by the plant operator, by taking into account daily forecasts of power production. Thus, this discharge spectrum can be considered as an additional boundary condition for the regulation algorithm, with the priority 5. The released discharge $Q_{out}(t_1)$ is

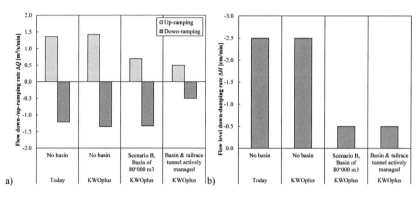

Figure 5. Flow regime characteristics for today's and the enhanced (KWO*plus*) Innertkirchen 1 HPP without retention volume, with one homogeneous retention volume of 80'000 m³ (Scenario B) and with active volume management for both basin and tailrace tunnel for winter conditions from 2009 to 2012: 95%-percentile of (a) flow up- and down-ramping rates immediately downstream of the outlet of the retention volume and (b) flow level down-ramping rate for flows below 8.1 m³/s for the gravel bars reach.

defined within this predicted discharge spectrum whenever possible, i.e. as long as no other boundary condition becomes determinant.

5.2 *Results*

The regulation performance is again given as the 95%-percentile of daily maximum values. Results reveal that even under operation and reliability constraints, flow up- and down-ramping rates can be considerably reduced compared to today's values or the rates expected after the enhancement of the Innertkirchen 1 HPP (KWO*plus*).

By actively managing the two storage volumes of the basin and the tailrace tunnel, the up-ramping rates can be reduced from 1.36 m³/s/min (today) and 1.43 m³/s/min (KWO*plus*) to 0.50 m³/s/min under optimum regulation parameters. The down-ramping rates are decreased to −0.50 m³/s/min for flows above 8.1 m³/s (Fig. 5a). For discharges below 8.1 m³/s, flow level down-ramping in the gravel bars reach is reduced to −0.5 cm/min, achieving the implemented threshold value (Fig. 5b).

The trailrace tunnel regulation is most effective for low turbine discharges when the unregulated volume occupied by the normal water depth is low and a considerable additional volume can be generated by closing the sector gates at the downstream end of the channel. This is especially important from an ecological point of view, as the flow level variations in the downstream gravel bars reach can be limited to ensure the conditions required for the brown trout during winter, when turbine discharges are generally low.

6 DISCUSSION

The increase of the turbine capacity of the Innertkirchen 1 HPP would increase the flow up- as well as down-ramping rates in the upper Aare River. A compensation volume, installed between the powerhouse and the release to the river, allows mitigation of these negative effects. To maximise its benefits, the operation rules have to focus on specific ecologically defined threshold values. In the given case, hydrological time series could have been produced for decision-making, taking into account different retention volumes and operation scenarios.

In a first step, the effect of one homogeneous retention volume on flow compensation has been studied. Even a small volume of 50'000 m³ allows a reduction of the up-ramping rate from 1.43 with KWO*plus* to 0.9 m³/s/min as well as of the flow level down-ramping rate in the ecologically relevant gravel bars reach from −2.5 to −0.5 cm/min, fulfilling the targets of the guidelines. However, a retention volume of 80'000 m³ has been defined as most suitable.

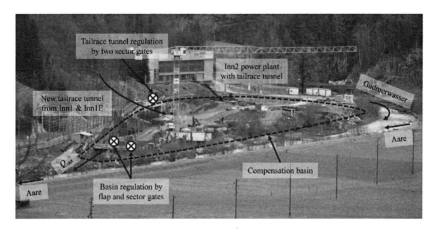

Figure 6. Construction site of the compensation basin in Innertkirchen (picture courtesy of KWO, January 2014).

According to the results of the preliminary study, this alternative results in up- and down-ramping rates of 0.7 and −1.33 m³/s/min as well as in a down-ramping rate in the ecologically relevant gravel bars reach of −0.5 cm/min.

In a second step, corresponding to the design phase of the enhancement project, the site conditions as well as additional system and operation requirements have been considered. The flow regulation downstream of the powerhouses is guaranteed by two retention volumes, namely the tailrace tunnel and the compensation basin, which are actively managed by gates. Under optimum regulation control, up- and down-ramping rates of 0.5 and −0.5 m³/s/min as well as a down-ramping rate of −0.5 cm/min in the gravel bars reach could be achieved for the given discharge series for winter conditions from 2009 to 2012. Up- and down-ramping rates could be improved by the developed design and the corresponding operation rules.

The chosen approach is straight forward. It focuses on the implementation of the mitigation measure. Target species as well as specific hydrological and morphological conditions of the Hasliaare River have been addressed. Several assumptions had to be made during the final design of the retention basin and tunnel, generating uncertainty regarding results. Future turbine operation is related to past winter conditions of 2009 to 2012, which may not fully correspond to future production process. The applied operation flexibility defined during detailed design should be able to address future changes in the schedule. Furthermore, future river restoration projects should consider the modified flow regime, avoiding dewatering of fish and its spawning ground. The system, as defined by the herein presented approach, is under construction now (Fig. 6). A monitoring system will allow assessing the performance and optimising system operation accordingly. Re-evaluation of the operation schedule has to be done continuously by addressing hydrological, plant operation, morphological and habitat conditions.

Nevertheless, flow regime mitigation is only successful with suitable river morphology and vice versa. Habitat simulations show that hydropeaking impact is strongly dependent on river morphology. Similar to many rivers in mountainous catchment areas, the Hasliaare River has undergone considerable anthropogenic changes. Construction mitigation measures, such as the compensation basin, would even show higher ecological performance for a naturally braided morphology. Thus, several river widening projects as well as instream improvements are under evaluation in the framework of flood management projects and KWO*plus*.

7 CONCLUSION

The applied method for flow restoration in the upper Aare River in Switzerland is presented, containing the definition of river specific habitat criteria, hydrological simulations as well as the detailed design of the retention basin. Flow restoration in Alpine streams affected by hydropower operations ask for specific indices for an appropriate assessment of the hydropeaking impact on aquatic habitat. For effective flow regime mitigation, restoration of the altered morphology is essential. The study may help to support the application of the Law on Water Protection for river restoration projects at existing and newly developed hydropower facilities in Alpine areas, showing beside conceptual approaches also realisation focused engineering.

The applied approach allows operators of hydropower plants, authorities or researchers to analyse impacted river systems, to design and rate ecologically and economically retention measures. Thus hydropeaking can be addressed in an optimal manner, as shown for one of the first hydropeaking retention basins in Switzerland.

REFERENCES

Baumann, P., Kirchhofer, A. & Schälchli, U. 2012. Sanierung Schwall-Sunk—Strategische Planung: Ein Modul der Vollzugshilfe Renaturierung der Gewässer. Report. Umwelt-Vollzug, Berne, Switzerland. In German.

Bunn, S.E. & Arthington, A.H. 2002. Basic principles and ecological consequences of altered flow regimes for aquatic biodiversity. *Environmental Management* 30(4): 492–507.

Moog, O. 1993. Quantification of daily peak hydropower effects on aquatic fauna and management to minimize environmental impacts. *Regulated Rivers: Research & Management* 8(1–2): 5–14.

Parasiewicz, P., Schmutz, S. & Moog, O. 1998. The effect of managed hydropower peaking on the physical habitat, benthos and fish fauna in the river Bregenzerach in Austria. *Fisheries Management and Ecology* 5: 403–417.

Person, E., Bieri, M., Peter, A. & Schleiss, A.J. 2014. Mitigation measures for fish habitat improvement in Alpine rivers affected by hydropower operations. *Ecohydrology* 7(2): 580–599.

Petts, G.E. 1984. Impounded rivers. Wiley, Chichester, UK.

Petts, G.E. & Amoros, C. 1996. Fluvial hydrosystems. Chapman and Hall, London, UK.

Poff, N.L., Allan, J.D., Bain, M.B., Karr, J.R., Prestegaard, K.L., Richter, B.D., Sparks, R.E. & Stromberg, J.C. 1997. The natural flow regime. *BioScience* 47(11): 769–784.

Schleiss, A. 2007. L'hydraulique suisse: Un grand potentiel de croissance par l'augmentation de la puissance. *Bulletin SEV/VSE* 07(2): 24–29. In French.

Schweizer, S., Neuner, J., Ursin, M., Tscholl, H. & Meyer, M. 2008. Ein intelligent gesteuertes Beruhigungsbecken zur Reduktion von künstlichen Pegelschwankungen in der Hasliaare. *Wasser Energie Luft* 100(3): 209–215. In German.

Schweizer, S., Meyer, M., Heuberger, N., Brechbühl, S. & Ursin, M. 2010. Zahlreiche gewässerökologische Untersuchungen im Oberhasli. *Wasser Energie Luft* 102(4): 289–300. In German.

Schweizer, S., Meyer, M., Wagner, T. & Weissmann, H.Z. 2012. Gewässerökologische Aufwertungen im Rahmen der Restwassersanierung und der Ausbauvorhaben an der Grimsel. *Wasser Energie Luft* 104(1): 30–39. In German.

Schweizer, S., Schmidlin, S., Tonolla, D., Büsser, P., Meyer, M., Monney, J., Schläppi, S. & Wächter, K. 2013a. Schwall/Sunk-Sanierung in der Hasliaare—Phase 1a: Gewässerökologische Bestandesaufnahme. *Wasser Energie Luft* 105(3): 191–199. In German.

Schweizer, S., Schmidlin, S., Tonolla, D., Büsser, P., Meyer, M., Monney, J., Schläppi, S., Schneider, M., Tuhtan, J. & Wächter, K. 2013b. Schwall/Sunk-Sanierung in der Hasliaare—Phase 1b: Ökologische Bewertung des Ist-Zustands anhand der 12 Indikatoren der aktuellen BAFU-Vollzugshilfe. *Wasser Energie Luft* 105(3): 200–207. In German.

Schweizer, S., Bieri, M., Tonolla, D., Monney, J., Rouge, M. & Stalder, P. 2013c. Schwall/Sunk-Sanierung in der Hasliaare—Phase 2a: Konstruktion repräsentativer Abflussganglinien für künftige Zustände. *Wasser Energie Luft* 105(4): 269–276. In German.

Schweizer, S., Schmidlin, S., Tonolla, D., Büsser, P., Maire, A., Meyer, M., Monney, J., Schläppi, S., Schneider, M., Theiler, Q., Tuhtan, J. & Wächter, K. 2013d. Schwall/Sunk-Sanierung in der Hasliaare—Phase 2b: Ökologische Bewertung von künftigen Zuständen. *Wasser Energie Luft* 105(4): 277–287. In German.

Young, P., Cech, J. & Thompson, L. 2011. Hydropower-related pulsed-flow impacts on stream fishes: A brief review, conceptual model, knowledge gaps and research needs. *Reviews in Fish Biology and Fisheries* 21(4): 713–731.

Swiss Competences in River Engineering and Restoration – Schleiss, Speerli & Pfammatter (Eds)
© *2014 Taylor & Francis Group, London, ISBN 978-1-138-02676-6*

Morphodynamic changes in a natural river confluence due to a hydropower modified flow regime

M. Leite Ribeiro
Stucky SA, Renens, Switzerland

S. Wampfler
Basel, Switzerland

A.J. Schleiss
Laboratory of Hydraulic Constructions (LCH), Ecole Polytechnique Fédérale de Lausanne (EPFL), Lausanne, Switzerland

ABSTRACT: River channel confluences form important morphological components of any river system. This article presents the results of an investigation on a natural river confluence in Switzerland. The aim of the field work performed at the confluence of the Sarine (main channel) and Gerine (tributary) Rivers near Fribourg (CH) is to analyze the interaction between morphological processes on the confluence and hydropower-affected hydrology of the main River. Important morphological changes have been observed over the last years, especially the deflection of the Sarine River and erosion of the left bank opposite of the tributary. Numerical calculations have shown that currently the minimal discharge of the Sarine River which transports the sediments annually deposited at the confluence is attained on average once each 2 years. Before the construction of the Rossens Dam, the minimum discharge was reached every year. This lack of yearly floods with sediment mobilization can explain the morphological changes occurred at the confluence since the construction of the dam.

1 INTRODUCTION

River confluences are the nodes of the fluvial network. They are zones where an important interaction between flow dynamics, sediment transport and bed morphology occurs (Best, 1988, Boyer et al., 2006, Rhoads et al. 2009). The interplay between sediment loads and the discharges in the tributary and the main channel appears to be the critical factor governing confluence morphology (Leite Ribeiro et al. 2012a,b).

Confluence hydrodynamic zones (CHZ), i.e., the zones around the confluence affected by the flow convergence (Kenworthy and Rhoads, 1995) are critical points with respect to the lateral and longitudinal connectivity of the network. It may represent biological hotspots in river networks (Benda et al., 2004; Rice et al., 2008) as in natural conditions they are typically characterized by high habitat heterogeneity, high variability in flow, sediment load, sediment size. These are requisites for high quality fluvial ecosystems.

This article presents a field investigation performed at the natural river confluence between the Sarine (main) and Gerine (tributary) Rivers near Fribourg (Switzerland). The morphology of the confluence has been affected by a significant change in the flow regime of the main river due to the implementation of a storage hydropower scheme some kilometers upstream. The objective of the study is to analyze the morphological changes occurred in this confluence since the construction of the Rossens Dam. The analysis has been done based on historical maps of the confluence, as well as a field investigation performed in 2008. The study has been completed by a simplified two-dimensional numerical study, where the objective was to determine the threshold conditions for the sediment transport at the confluence.

2 SARINE-GERINE CONFLUENCE

The confluence of the Sarine (main) and Gerine (tributary) Rivers is illustrated in Figure 1. This confluence is one of the rare confluences in Switzerland that is still in its natural condition, i.e. without any engineering intervention on the CHZ.

A three-dimensional representation of the Gerine and the Sarine Rivers upstream of their confluence is represented in Figure 2.

2.1 The Gerine River

The source of the Gerine River (drainage area of 79 km²) is on the Massif de la Berra at elevation 1500 masl, around 20 km upstream of the confluence. The Gerine is a torrential river, with bed slopes varying from 1.6% at the mouth to 46% at Berra. The streams feeding the Gerine are very short and their slopes are always higher than 6%.

Along its course, the Gerine River is subjected to alternating deficit and excess of sediments. In the upstream reach, the river behaves as a torrent with an erosion regime. There is an

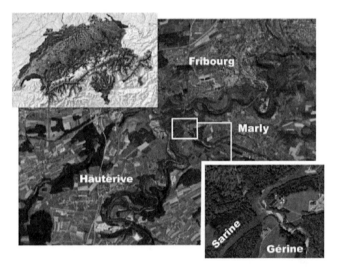

Figure 1.   Location of the Sarine-Gerine confluence in Switzerland.

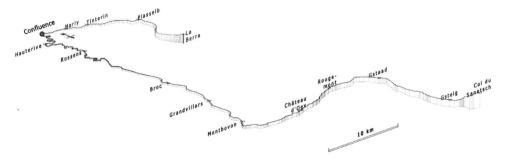

Figure 2.   Three-dimensional representation of the Sarine and Gerine Rivers upstream of their confluence.

192

erosion zone at Plasselblund that is followed by an alluvial cone at Sagenboden. The sequence erosion zone—alluvial cone is repeated between Tinterin and the confluence. The annual volume of sediments transported to the confluence is around 40'000 m³ (Wampfler, 2008).

2.2 *The Sarine River*

The Sarine River rises at Sanetsch, at elevation 2252 masl. The length of the river until the confluence with the Gerine is about 85 km (surface of the river basin 981 km²). The river has a very irregular bed slope, varying from a near-vertical profile at the Sanetsch Dam and a bed slope of 0.45% near the confluence.

There are several dams constructed along its course, which influences the sediment transport along the river. The largest one is the Rossens Dam, located around 15 km upstream of the confluence. The artificial lake created by the Rossens Dam works as huge sediment retaining reservoir. Downstream of the dam, the sediment input and transport is very limited until the confluence.

2.3 *Hydrology*

The flood peak discharges of the confluents are presented in Table 1. For the Gerine River, the values are based on the gauging station located at Marly some kilometers upstream of the confluence (drainage area at the station = 77.6 km²). In the case of the Sarine River, a hydrological analysis has been made based on the fluvial gauging stations located at Broc (around 32 km upstream of the confluence), Fribourg (around 10 km downstream) and Laupen (around 28 km downstream).

2.4 *Rossens Hydropower Scheme*

Figure 3 presents an overview of the Rossens Hydropower Scheme. The lake created by the Rossens Dam (arch dam 83 m high) is called Lake of Gruyère and has a surface of 9.4 km² and a volume of 200 million m³. This is the longest artificial lake in Switzerland.

Table 1. Flood discharges of the Gerine and Sarine Rivers at the confluence.

T (years)	1	2	5	10	20	50	100	200
$Q \ (m^3/s)$								
Gerine	56	68	82	92	103	117	127	138
Sarine	84	250	352	414	464	524	565	602

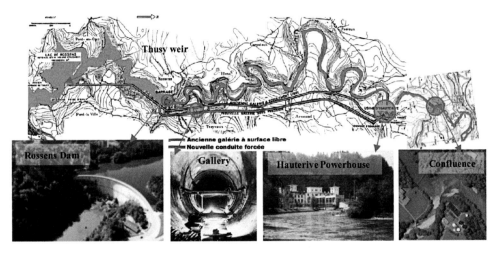

Figure 3. Rossens Hydropower Scheme.

193

The Hauterive Powerhouse is located between the Dam (13.5 km upstream) and the confluence (1.5 km downstream). The plant is connected with the lake by a circular gallery approximately 6 km long constructed at the right bank. The maximum discharge through the gallery is 75 m³/s. The short-circuit reach between the intake and the powerhouse is called Little Sarine. It is supplied by a residual discharge of 3.5 m³/s during the summer and 2.5 m³/s during the winter (Mivelaz, 2005). The Little Sarine is not subjected to the hydropeaking effects and therefore this reach plays a vital role in the reproduction of aquatic fauna.

The Hauterive powerhouse was commissioned in 1902 and initially turbined water from the Thusy Dam, located around 3.2 km upstream of Rossens. Since the construction of the Rossens Dam in 1948, the hydrological regime of the Sarine has changed significantly due to the hydropeaking regime (Fig. 4) and flood routing by the artificial lake upstream of the confluence.

The number of floods and flood magnitudes reduced considerably at the confluence. The average annual peak flow decreased by 120 m³/s for the gauging station in Fribourg (Fig. 5).

Figure 4.   Effect of the hydropeaking on the confluence.

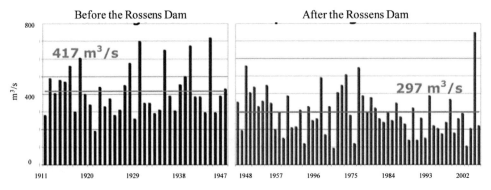

Figure 5.   Annual peaks of the Sarine River from 1911 to 2007 in Fribourg (Swiss Federal Office for the Environment, 2007).

194

3 MORPHOLOGICAL EVOLUTION OF THE CONFLUENCE FROM 1945 TO 2005

The morphological evolution of the confluence has been analyzed by means of historical and recent maps (Fig. 6). The historical Siegfried Map elaborated in 1945 illustrates the situation before the construction of the Rossens Dam, whereas the Swiss National Maps elaborated in 1974, 1992, 1998 and 2005 present the morphological evolution of the confluence from the construction of the dam. The overlay contours reveals the changes between each period.

It can be noted that the operation of the Rossens Hydropower scheme affected considerably the dynamics of the confluence. Before Rossens Dam, the Sarine regime predominated over the Gerine regime, whereas from 1948, this situation has been reversed.

Between 1945 and 1974 the left bank of the Sarine was completely eroded downstream of the confluence, forming a vertical rocky cliff. Major morphological changes are also visible on the floodplain. The fact that the Gerine course has changed indicates that large floods took place during this period (29 years). The construction of a new road bridge crossing the Gerine forced the passage of this river to the left bank.

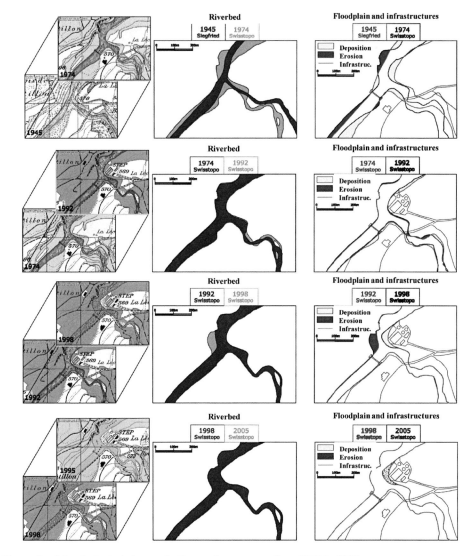

Figure 6.   Morphological changes in the confluence zone from 1945 to 2005.

From 1974 to 1992, the confluence was not subject to significant changes. The erosion of the left bank has slightly increased. The flood of the Sarine in 1977 ($Q = 550$ m³/s) had no significant impact on the morphology of the confluence.

The morphological change observed between 1992 and 1998 is very marked. A large area has been eroded on the left bank showing that important differences can occur over a limited time. Between 1992 and 1998, the Sarine has not experienced important floods, able to significantly alter the morphology of the confluence.

The erosion of the left bank between 1992 and 1998 is probably due to floods in the Gerine River occurred in 1995 ($Q_{max} = 101.8$ m³/s) and on 1997 ($Q_{max} = 147.7$ m³/s). The 1997's flood has a return period larger than 200 years.

Between 1998 and 2005, the morphology of the confluence remained unchanged. The 2005's flood occurred in the Sarine has resulted in a maximum of 680 m³/s. The flow of the Sarine was probably sufficient to evacuate the material transported by the Gerine.

4   TWO-DIMENSIONAL NUMERICAL SIMULATIONS

Simplified hydrodynamic numerical simulations have been performed with CCHE2D, a two-dimensional depth-averaged model developed by the National Center for Computational Hydroscience and Engineering of the University of Mississipi, USA. The objective of the simulations performed under steady flow conditions with fixed bed and without sediment transport is to determine the minimum discharge of the Sarine River, which is able to transport the sediments deposited at the confluence.

4.1   Basic data

4.1.1   Bathymetry
Most of the bathymetry of the confluence used for the simulations has been surveyed in 2000 (Ribi, 2000). This bathymetry has been completed by a simplified survey of the confluence bar performed in 2008. Figure 7 presents the bathymetry used for the simulations.

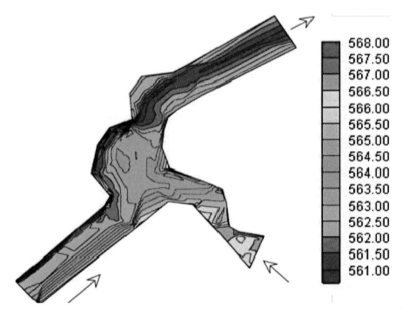

	568.00
	567.50
	567.00
	566.50
	566.00
	565.50
	565.00
	564.50
	564.00
	563.50
	563.00
	562.50
	562.00
	561.50
	561.00

Figure 7.   Bathymetry of the confluence used for the numerical simulations. Color bars represent the bed elevation (masl).

4.1.2 Grain size characteristics

The grain size characteristics of the Sarine and the Gerine Rivers are presented in Table 2. The characteristics of each river have been determined following the Linear Sampling Method proposed by Fehr (1987).

4.2 Methodology

In order to assess the sediment transport behaviour at the confluence, 7 scenarios with different combinatios between the Sarine and Gerine discharges have been considered (Table 3). The analysis has been performed by comparing the driving forces of particle motion (shear stress) to the resisting forces that would make it stationary (particle density and size). For that comparison, the Shields Diagram (Shields, 1936) has been used.

4.3 Results

In the confluence zone, the dimensionless diameter d_* calculated according to Shields (Equation 1) is equal to 1'264. It means that the flow is in the rough turbulent regime and the critical dimensionless shear stress t_{*cr} can be adopted as 0.047. Therefore, the critical shear stress τ_{cr} (Equation 2) is around 46 N/m².

$$d_* = d\left(\frac{\rho_s - \rho}{\rho} * \frac{g}{v^2}\right)^{\frac{1}{3}} \tag{1}$$

where $d*$ = dimensionless diameter, d (m) = characteristic particle diameter of the sediment, ρ_s (kg/m³) = density of the sediment, ρ (kg/m³) = density of the water, g (m/s²) = acceleration due to gravity and v (m²/s) = kinematic viscosity.

$$\tau_{cr} = t_*(\gamma_s - \gamma)d \tag{2}$$

where τ_{cr} = critical shear stress (N/m²), t_* = critical dimensionless shear stress, γ_s (N/m³) = specific weight of the sediment, γ (N/m³) = specific weight of the water and d (m) = characteristic particle diameter of the sediment.

Table 2. Grain size characteristics of the sediments found in the Sarine and the Gerine Rivers.

River	d_m (cm)	d_{50} (cm)	d_{90} (cm)	Source
Gerine	9	6	22	Jaeggi, 1999
Sarine	5	3	10	Wampfler, 2008

Table 3. Simulated hydraulic scenarios.

Scenario	Q_{Sarine}	Q_{Gerine}
1	$Q_1 = 84$ m³/s	$Q_1 = 56$ m³/s
2	$Q = 100$ m³/s	
3	$Q = 125$ m³/s	
4	$Q = 150$ m³/s	$Q = 2$ m³/s
5	$Q = 175$ m³/s	
6	$Q = 200$ m³/s	
7	$Q_2 = 225$ m³/s	

From Figure 8, it can be observed that the critical shear stress is already exceeded for annual floods in the Gerine River upstream of the confluence. Therefore, it suggests that the sediment transport at the Gerine starts from events with peak discharges lower than those corresponding to the annual flood.

By analyzing the influence of the Sarine's discharge on the transport capacity over the confluence bar (line A-B on Figure 9, left), it has been found that critical shear stress of 46 N/m² is exceed for peak discharges of the Sarine of around 175 m³/s. This corresponds to approximately a 2-year return period flood ($Q_2 = 250$ m³/s) after the construction of the Rossens Dam. The spatial distribution of the shear stresses for the scenario 5: $Q_{Sarine} = 175$ m³/s and $Q_{Gerine} = 2$ m³/s is presented in Figure 9 (right).

Therefore, it seems that the confluence is currently not in equilibrium from the sediment transport point view. There is an imbalance between the frequency of the sediment transport in the Gerine (<1 year) and in the Sarine (~2 years). This imbalance is supposed to be responsible for the main morphological changes occurred in the confluence since the construction of the Dam. The sediments annually transported by the Gerine are deposited in the confluence bar. As it takes on average two years so that the confluence bar can be washed, the obstacle

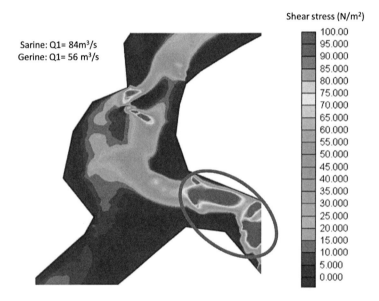

Figure 8.   Shear stresses calculated for annual floods at Sarine and at Gerine Rivers.

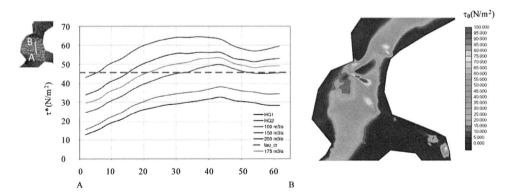

Figure 9.   Shear stresses at the confluence bar over the line A-B (left) and spatial distribution of the shear stresses (N/m²) on the confluence zone for the scenario 5 ($Q_{Sarine} = 175$ m³/s / $Q_{Gerine} = 2$ m³/s).

198

formed by the bar imposes a course change of the Sarine River bed. For comparison, the discharge of 175 m³/s was exceeded every year before the construction of the Rossens Dam.

5 CONCLUSIONS

The confluence of two channels, each of them having independent flow and sediment discharge regimes creates complex erosional and depositional conditions. Changing these input conditions can provoke important changes in channel morphology at the confluence.

The Rossens Hydrolelectric Scheme, constructed in 1948 significantly influences the hydrology of the Sarine River. This is a key element for explaining the morphological processes, especially the large erosion at the left bank occurred at the confluence between the Sarine and the Gerine Rivers over the last years.

Simplified numerical simulations have shown that the minimum discharge of the Sarine to transport the material annually deposited at the confluence by the Gerine is around 175 m³/s. This discharge is slightly lower than the peak value of the 2-year return period flood, determined after the construction of the Rossens Dam. The imbalance between the sediment transport frequencies on the two confluents has been responsible for the morphological changes occurred in the confluence since the construction of the dam.

REFERENCES

Benda, L., Poff, N.L., Miller, D., Dunne, T., Reeves, G., Pess, G. and Pollock, M. 2004. The network dynamics hypothesis: How channel networks structure riverine habitats, *BioScience*, 54, 413–427.
Best, J.L. 1988. Sediment transport and bed morphology at river channel confluences, *Sedimentology*, 35, 481–498.
Boyer, C., Roy, A.G. and Best, J. 2006. Dynamics of a river channel confluence with discordant beds: Flow turbulence, bed load sediment transport, and bed morphology. *Journal of Geophysical Research*, 111: F04007 1_22.
Fehr, R. 1987. Einfache bestimmung der korngrössenverteilung von geschiebematerial mit Hilfe der Linienzahlanalyse. *Schweizer Ingenieur und Architekt*, 38:1103–1109, 1987. (in German).
Kenworthy, S.T., and Rhoads, B.L. 1995. Hydrologic control of spatial patterns of suspended sediment concentration at a stream confluence, *Journal of Hydrology*, 168, 251–263.
Leite Ribeiro, M., Blanckaert, K., Roy, A.G. and Schleiss, A.J. 2012a. Flow and sediment dynamics in channel confluences. *Journal of Geophysical Research—Earth Sciences*, doi:10.1029/2011 JF002171.
Leite Ribeiro, M., Blanckaert, K., Roy, A.G. and Schleiss, A.J. 2012b. Hydromorphological implications of local tributary widening for river rehabilitation. *Water Resources Research*, 48(10).
Mivelaz, L. 2005. Augmentation du débit de dotation de la Petite Sarine en aval du barrage de Rossens, 2005. Entreprises Electriques Fribourgeoises, Direction Production Energie. Available at *http:///lch. epfl.ch/.*
Rhoads, B.L., Riley, J.D., and Mayer, D.R. 2009. Response of bed morphology and bed material texture to hydrological conditions at an asymmetrical stream confluence, *Geomorphology*, 109, 161–173.
Ribi, J-M. 2000. Aménagement de la Gérine—Mesures de protection contre les crues. *Ribi SA*. Report not published. (in French).
Rice, S.P., Kiffney, P., Greene, C. and Pess, G.R. 2008. The Ecological Importance of Tributaries and Confluences, 209–242 pp., *John Wiley & Sons, Ltd.*
Shields, I.A., 1936, Anwendung der Ähnlichkeitsmechanik und der Turbulenzforschung auf die Gescheibebewegung, *Mitt. Preuss Ver.-Anst.*, 26 (in German).
Wampfler, S. 2008. Morphologie des confluences naturelles aménagées—Etude de terrain d'une confluence naturelle et étude expérimentale sur une confluence en modèle, edited, p. 211, Master Project. *Laboratory of Hydraulic Constructions (LCH)—Ecole Polytechnique Fédérale de Lausanne (EPFL)* (in French).

Author index